Introduction to Process Plant Projects

Introduction to Process Plant Projects

H. Selcuk Agca
Giancarlo Cotone

CRC Press
Taylor & Francis Group
Boca Raton London New York

CRC Press is an imprint of the
Taylor & Francis Group, an **informa** business

CRC Press
Taylor & Francis Group
6000 Broken Sound Parkway NW, Suite 300
Boca Raton, FL 33487-2742

First issued in paperback 2020

ISBN-13: 978-1-138-60668-5 (hbk)
ISBN-13: 978-0-367-73358-2 (pbk)

Library of Congress Cataloging-in-Publication Data

Names: Agca, H. Selcuk, author. | Cotone, Giancarlo, author.
Title: Introduction to process plant projects / H. Selcuk Agca and Giancarlo Cotone.
Description: Boca Raton: CRC Press; Taylor & Francis, [2019] |
Includes bibliographical references and index.
Identifiers: LCCN 2018018710 (print) | LCCN 2018025444 (ebook) |
ISBN 9780429882616 (Adobe PDF) | ISBN 9780429882609 (ePub) |
ISBN 9780429882593 (Mobipocket) | ISBN 9781138606685 (hardback) |
ISBN 9780429466762 (ebook)
Subjects: LCSH: Project management. | Process control. | Industrial procurement. |
Construction projects—Management. | Engineering—Management.
Classification: LCC HD69.P75 (ebook) | LCC HD69.P75 A39 2019 (print) |
DDC 658.5/1—dc23
LC record available at https://lccn.loc.gov/2018018710

Contents

PART II   Launching the Project

PART III   *Developing the Project*

PART IV   *Later Stages of Project Development*

PART V   *Project Completion*

PART VI   *Other Functions with Top Priority*

Preface

This book aims at highlighting the main aspects of implementing industrial plant projects to those readers who have little or no familiarity with execution of such complicated and multidisciplinary activities. In other words, the objective is set out to provide a kind of roadmap for new starters, like young engineers, new graduates, or engineering students who are either just about to be stepping in or checking out such opportunities for future careers in process industries. Special emphasis is given to process plants due to the fact that these are usually more complicated than other types of industrial plants, yet both in fact follow very similar steps along project implementation practices.

The book is an outcome of the authors' experience on process facility projects as well as feedback received during seminars and lectures they have given to engineering students and young engineers. It intends to address the target readers in a simple and orderly manner to provide ease in getting connected and follow-up. For such purposes, the book is arranged in "parts" which consist of a number of "chapters" based on the following:

- The four chapters in "Part 1 – Basic definitions" make fundamental descriptions of the terms, concepts, and parties involved in project execution environments, with the aim to prepare the reader for going through the subsequent parts.
- The chapters from "Part 2 – Launching the project" through "Part 5 – Project completion" are devised to form a comprehensive and sequential explanation of the entire project stages.
- The last part, "Part 6 – Other functions with top priority," mainly deals with the functions that ensure all project activities are started and progressed in line with certain crucial requirements for health–safety–environment, quality management, contract administration, etc.
- In general, the first chapters of each part are intended to make an introduction for a more detailed description of those topics in the next chapters of the same part.
- The appendices at the end of the book contain relevant auxiliary information to support the material presented in the main body of the text. They intend to expand the readers' knowledge on such related fields.

The core outlook of the book is selected to be from an engineering contractor's perspective that performs either on E (engineering), EPCm (engineering–procurement–construction management), or EPC (engineering, procurement, and construction) basis. The project initiation and follow-up steps by the investors/ owners are also mentioned, though a bit more briefly compared to the case for contractors.

Generally, emphasis is put on home office services, as these form the essentials for project execution activities that require to be discovered by young readers, with clear answers to "who," "what," "how," "to which extent," "when," and "why." It is deemed that the target readers certainly need to have a good idea about the foregoing concepts, before they can start learning the other related stages and functions of the main project execution theme, like proposals, contract management, project strategy, construction management, etc., which are placed at the back end of the text, accordingly. Moreover, regardless of whether a young engineer is employed by an investing owner or an engineering contractor and assigned to a project, he or she will most probably be starting the job by getting involved directly with the scope and project execution activities through home office services. Such considerations resulted in having the front-end chapters of the book dedicated to execution of projects, just kicked off to proceed with home office services. In the world of process plant projects, activities prior to real start of project execution, like feasibility studies/reviews, financing, investment decisions, and contract negotiations are normally handled by clients at much higher staff levels (e.g., senior engineers, managers) than juniors. Hence, such activities are mentioned in the book either briefly or covered in detail in the appendices for the readers who are interested to know more on these topics.

It is also worthwhile mentioning that the book is set to tell the story of industrial facility projects (whether small or large, revamp or new plant, petroleum, chemical, power or food, etc.) in line with the techniques generally followed by global clients and global engineering companies. Thus, it is likely that some other clients and engineering companies may be performing their projects a little bit differently from what is described in this book with regard to some of the activities and documentation mentioned.

H. Selcuk Agca
Giancarlo Cotone

Acknowledgments

We would like to thank our friends for their encouragement in writing this book. Our special thanks are for the following peers:

- Prof. Dr. Mustafa Ozilgen for his valuable support and suggestions about publishing the book.
- Luigi Bressan for his encouragement, suggestions, and peer review. Moreover, it was Luigi who introduced the authors of this book to each other.
- Nevzat Tunay for his comments on HSE.
- Mesut Inan for his comments on project control.
- Yavuz Serdar for drafting the sample engineering drawings.
- Bulent Karaca for the 3D plant model.
- Prof. Dr. Nurcan Bac, Halil Semerci, James Woolley, and Tim Zboya for their peer reviews before publishing.

A warm thanks goes to our families for their support and patience while we worked long hours during most weekday nights and entire weekends on this endeavor.

About the Authors

Giancarlo Cotone is an Italian chemical engineer who obtained his degree from the University of Roma (Italy) in 1971. His entire career has unfolded in the world of engineering and construction in the fields of oil and gas, refining, power, and petrochemical. He has worked for first class contractors such as Snamprogetti and Foster Wheeler, covering a variety of roles, from process engineering to feasibility studies, from international subcontracting to contract administration, from proposals to sales and marketing. His last assignment was as a senior vice president, Sales and Marketing for the Middle East.

Since his retirement in 2009, he cooperated with medium- and small-size Italian contractors on sales and marketing and strategic positioning issues, especially in the Middle East. He is also a teacher on contract administration and procurement subjects, both for a prestigious Italian business school and for private institutions.

Giancarlo consistently travelled around the world for business, and spent over 13 years of his life overseas, in France, Venezuela, the United Kingdom, Bahrain, and Saudi Arabia.

He is married and has a son – an engineer as well – who gave him two wonderful grandchildren. In his free time, he loves gardening and has purchased an olive grove in Tuscany where he spends a considerable portion of his time tending to olive trees and his vegetable garden.

H. Selcuk Agca has BS and MS degrees in chemical engineering from the Middle East University of Ankara, Turkey. While working as a teaching assistant at the university, he also started PhD studies, but decided shortly to move to the field of professional engineering on process plants. He worked in the Istanbul office of Foster Wheeler for about 25 years, holding the position of contract operations director during a notable portion of this period.

Since 2010, Selcuk has been giving lectures at Yeditepe University of Istanbul on a part-time basis. He has mainly taught process plant design to senior students of the chemical engineering and food engineering departments. During the same period, he has taken part in accreditation of the engineering departments of Turkish universities as an assessor for Mudek, the accreditation association for engineering education in Turkey. Meanwhile, he has worked as a consultant on process plant projects. He currently works as a consultant to Tekfen Construction, Inc. on global EPC and EPCm proposals.

Selcuk is married and has a daughter. His wife and his daughter are engineers, too. His main hobbies are music and cooking. He collects bass guitars of prestigious brands and models.

Part I

Basic Definitions

1 Preamble

Implementation of industrial plant projects requires making use of a number of different types of specific expertise in a variety of specialty fields, each calling for lifelong accumulation of knowledge and on-the-job experience. Industrial plants (and more specifically the process plants) are one of the most sophisticated systems created and continuously upgraded along the booming technological advances, started virtually with the industrial revolution, for conversion of raw materials into other materials.

The following parts and chapters of the book are prepared to deliver essential information regarding entire project life, including major project phases, specialty fields of project functions and project team positions taking part in the execution.

So, let us start with defining briefly some of the important basic concepts that require having a common understanding before getting inside the professional life of project execution.

WHAT IS A "PROJECT"?

Definitions for the word "project" in dictionaries are sometimes quite superficial, such as "a course of actions intended or considered possible," "a systematic planned undertaking," and "a piece of work, often involving many people, that is planned and organized systematically."

Such descriptions are not incorrect; however, a far more profound description is still needed in order to professionally reflect what a "project" really means.

Giving some examples from various different types of projects can help notably before trying to make up a clearer definition for this term. Here are some below:

- Preparing new computer software
- Writing a movie script, shooting a new motion picture
- Composing songs, recording a new music album
- Designing and manufacturing prototype of a new product
- Exploring new oil or mining fields
- Constructing a new or revamping an existing production facility (this book will focus on this type of projects)

Many other examples can of course be added to the above list. However, let us now go back to our main intention of describing what a project really is. For this purpose, let us start listing specific features of projects. A project

- Has a particular purpose and certain requirements.
- Has some cost that is to be estimated and budgeted.
- Requires to be planned and organized in sufficient detail including setting up a time schedule.

3

- Needs certain resources including manpower.
- Is expected to give a benefit in return, at the end.
- Requires different trades and expertise to be integrated and coordinated.
- Has to be monitored and controlled with respect to time/progress, cost, resources, and quality.
- Has to take care for human health, safety, and environment if the implementation and/or the final product poses risks for these (i.e., often when the project involves manufacturing, construction, and plant operation).
- Does not yield any standard product (mass production/assembly line products) or regular/routine services.
- Has activities, inputs, and outputs, almost all of which are specifically tailored to fit the purpose.
- Has no other output (product or service) that is exactly identical to any previous one.
- Has a life, i.e., starts, develops, and ends.

The foregoing outline now looks more comprehensive with regard to explaining the term. Please note that especially the last four characteristics above differentiate a project from other type of business activities.

WHAT DOES THE WORD "INDUSTRY" REFER TO?

Industry is an economic sector dealing with production of goods and/or services. There are a number of different types of classifications for industry depending on the objective of the classification. Several different classification systems are used in market research, finance, and statistics, such as ICB (Industry Classification Benchmark), GICS (Global Industry Classification Standard), NAICS (North American Industry Classification System), and ISIC (International Standard Industrial Classification). Anyway, key industrial economic sectors can broadly be identified as follows:

Primary sector: Raw material extraction including mining, petroleum extraction, and agriculture
Secondary sector: Manufacturing, refining, construction
Tertiary sector: Services, distribution of manufactured goods
Quaternary sector: Technological activities and services including research and development (R&D) and computer programming

For the material covered in the following sections of this book, the focus will obviously be on the secondary sector in which the industries involved are further identified by the nature of the products, as chemical industry, power industry, etc.

WHAT HAS BEEN THE IMPACT OF "INDUSTRIAL REVOLUTION" ON HUMAN LIFE?

Industrial revolution is the age that started within the second half of the 18th century. Before this era, Western Europe had already been inventing some sort of complicated machinery and tools since long ago, leading to developments in transportation and communication by introducing faster ships, printing facilities, etc.; however, the economy had still been based on agriculture depending very highly on manpower and manual handcrafting, traditionally carried out either at home or in small workshops. By the second half of the 18th century, a number of innovations in textiles, steam power, and iron founding kicked off the first industrial revolution, basically in Britain, and then spread throughout the Western Europe and North America. Fast expansion of trading due to substantial improvements in transportation and communication enhanced this great change to trigger industrialization. Such changes were also bringing in new ways in organizing business and labor. Scientific knowledge was started to be applied to production methodologies and business practices on an increasing scale. Start of the mass production has led to increased production capacities, higher efficiencies, and lower prices.

Almost a century after the first one, the second industrial revolution arose as an American contribution to what had been achieved that far. The main invention was the development of "continuous" manufacturing processes, started mainly with flour milling, cigarette production, and canned food manufacturing. Development of American manufacturing technologies gradually spread out to other industries including iron-steel making, distilling, refining, etc., leading to the second industrial revolution that is regarded to commence by 1850s. Hence, while Britain had been the epicenter of the first revolution, the second one took place most powerfully in the United States of America.

The era is called the industrial revolution because it notably and quite rapidly changed the economy, human life, and hence, the society by way of introducing the following:

- New forms of basic materials (iron and steel, chemicals, etc.)
- New power sources (steam engines, internal combustion engines, petroleum, electricity, etc.)
- Invention of new machinery (like various types of spinning machines used in textiles)
- New organization of work (known as factory system)
- Advances in transportation (steamships, automobiles, airplanes, etc.)
- Developments in communication systems (telegraph, radio, etc.)
- Application of scientific knowledge to production methodologies

The industrial revolution is a major change in human history in having transformed the agricultural societies into the modern industrial societies.

WHAT DOES "PROCESS" MEAN?

"Process" simply means a series of operations that take place in a pre-established order. In our case, it is to treat something in order to produce something else with regard to physical and/or chemical properties. In other words, the word "process" refers to treating raw materials through a series of operations and sometimes, reactions for obtaining products.

WHO ARE "PROCESS ENGINEERS"?

Process engineers are the persons who deal with engineering, design, checking, and controlling industrial processes. They are mostly chemical engineers (or food engineers for food plants) that make use of the principles of mainly the momentum transfer/fluid flow, heat transfer, mass transfer/separation processes, reaction kinetics, process control, and thermodynamics in addition to the advances in process technology. For few of the processes, like fuel-fired power generation, some other engineers, like mechanical engineers, may take this position since operation of such plants does not require any notable background on mass transfer and reaction kinetics. The necessary main principles to follow become the ones that are commonly studied in both chemical and mechanical engineering, such as fluid flow, heat transfer, and thermodynamics.

As some reference has just been made to chemical engineering, it may be a good idea to clarify at this point the difference between chemistry and chemical engineering. A chemical engineer is basically an engineer who has some more knowledge on the science of chemistry than other engineers, but never as much as chemists and so, he or she rather deals with the industrial applications of chemical processes having a broader background on abovementioned engineering topics as well as economics and management.

Before getting into details of interrelated and integrated group of activities for industrial plant projects, different types of process plants and the sections that commonly exist in such plants will be briefly mentioned in the following paragraphs.

WHAT ARE "PROCESS PLANTS"?

Manufacturing industries are often identified with the type of main products (or processes), which require often quite complicated plants composed of many items of equipment, extensive and complicated configuration of piping and cables, sophisticated instrumentation and automation, huge amounts of structures, big network of underground facilities, etc. In certain types of plants like petroleum refineries and chemical/petrochemical complexes, the sophistication, complexity, and huge quantities involved are usually much above the others.

A listing of process plants according to the type of main process or main product can be made as follows:

- Oil and gas
- Chemical
- Petrochemical
- Power (fuel fired)
- Pharmaceutical
- Food
- Miscellaneous other

The classification of production industries is often further detailed, still to address a group of plant types. As an example, the chemical industries cover the following subgroups, even if the production scheme may be totally different within each such subgroup:

- Industrial gases
- Glass, cement, and ceramics
- Chlorine-alkali products
- Fertilizers
- Surface-coating chemicals
- Detergents, soaps, and personal care products
- Paper and pulp
- Polymers, synthetic fibers, and plastics
- Miscellaneous other chemical products

Similar to classifying process plants, projects are also named generically according to the industry or the main product, as a refinery project, power plant project, pharmaceutical plant project, etc. However, it should be noted that dealing with any of the above groups does not show major differences with respect to project implementation methodologies. So, in this book, almost no emphasis will be needed for the type of product or industry while describing the implementation of process plant projects.

Categorization of process plant projects is as well made with regard to project scope, or to the extent of the work, such as the following:

- Completely new (grass roots or green field) plant composed of all necessary facilities on a new site
- Revamping an existing plant for increasing capacity or efficiency (debottlenecking) or for modernization (e.g., upgrading equipment or instrumentation/ automation)
- One or more new units on an existing plant site (brownfield projects)

A brief review of sections of an established plant may help visualizing what kind of scope an industrial plant project may potentially cover. The sections that make up a process plant are most commonly the following:

- Process unit(s)
- Raw material receipt and storage
- Product storage and dispatch
- Utilities/off-sites and interconnecting
- Automation and control systems
- Electrical systems
- Drain/sewer systems and waste treatment
- Safety, security, and protection systems
- Buildings, structures, and roads

Process Unit(s)

Process units are onsite units which consist of a variety of major and minor process equipment items as well as other supplementary systems like electrical equipment, concrete and steel structures, field instrumentation, cabling in conduits or on cable trays, etc. Liquid and gas processing units most commonly consist of, as major equipment, heat exchangers, tanks, reactors, furnaces, pressure vessels, towers, pumps, compressors, turbines, various packaged systems, etc., with extensive onsite piping runs and connections having a variety of on-line appurtenances.

Solid processing (solid–liquid or solid–gas) plants or such parts of process plants additionally contain some other types of major equipment, mostly for conveying and storing solids, like conveyors, elevators, bunkers, chutes, rotary feeders, cyclone separators, dust collection filters as well as crushers/grinders, kilns, etc.

Configuration of a simple production unit indicating the functional blocks is shown in the generic block flow diagram in Figure 1.1. Each block may have been made up of multiple equipment items. The blocks for preparation of feed and preparation for separation involve mostly temperature and/or pressure adjustment, so may consist of heat exchangers, furnaces, pumps, compressors, or pressure-reducing stations. Dashed lines shown for streams and blocks mean such blocks and connections may not exist, depending on the processed materials and products. The recycle shown, if exists, stands for the back feeding of the unreacted raw material to the front end of the process in order to increase the conversion and save loss of the valuable raw materials. In some cases, one or more of the blocks may be repeated (e.g., another reaction step after separation). Production often requires certain utilities (such as steam and cooling water) shown in the figure with return streams (e.g., condensate for steam, cooling water return for cooling water) in dotted lines. Similarly, effluents generated during the production in various steps are collected and charged to waste treatment to be released to the environment after treated to allowable specifications.

The following are the most common non-process sections of an established plant.

RAW MATERIAL RECEIPT AND STORAGE

The characteristics of this section depend on several factors including the state (gas/liquid or solid), chemical and physical properties, the process needs, and the

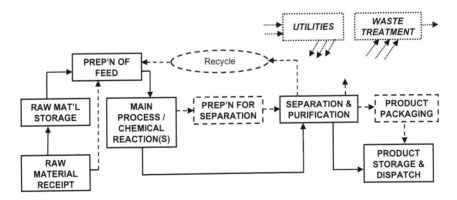

FIGURE 1.1   Anatomy of a simple process unit in generic block diagram form.

quantities concerned. Solids in relatively large quantities are received at the plant generally as bulks via trucks, railroad cars, ships, or conveyor systems and usually stock piled at an open storage site or dumped into big bunkers. Liquids and gases in large quantities are received through pipelines, unloading arms and piping manifolds on jetties, tanker trucks, or railroad cars. Liquids are stored mostly in atmospheric tanks, while gases are charged into fixed or telescopic drums. All these raw materials receipt and dispatch sections of plants are equipped with metering or weighing systems. Raw materials in relatively small quantities are received in containers, cylinders, bottles, boxes, etc. and stored mostly in enclosed warehouses unless material properties dictate some other ways of storage.

PRODUCT STORAGE AND DISPATCH

Storage and dispatch sections for products consist of similar installations as in raw material receipt and storage sections. Since some products are packaged for market requirements, those dispatch systems include also packaging equipment.

UTILITIES/OFF-SITES AND INTERCONNECTING

These are the non-process (off-site) units needed for running the main processes of the plant, from outside of the battery limits of process units. Some of these utilities are also used to serve other non-process purposes, like chilled water for air conditioning of buildings, compressed air for workshop tools, etc. Types, capacities, and specifications are mostly dictated by the processes. An industrial plant usually possesses the below listed utilities:

- Water (raw, filtered, potable, soft, and/or demineralized water)
- Cooling (tower) water
- Chilled water
- Steam and condensate, high pressure (HP), medium pressure (MP), and low pressure (LP)
- Hot oil
- Fuel (gas, oil, coal, waste)
- Compressed air (plant and instrument air)
- Inert gas (usually nitrogen)
- Electrical power (also classified under electrical systems below)
- Other (brine, tracing solvent, molten salt, refrigeration, directly received and returned cooling water from river, lake, or sea, etc.)

Interconnecting sections of a plant include long runs of piping and cabling between separate units of a plant in the form of the following:

- Pipe ways and/or pipe racks
- Cable racks, cable trays, and cable ladders
- Piping and cables either directly buried or placed in closed trenches

Cabling is either arranged to run separately in closed trenches or in pipe ways (pipe bundles with supports in large open trenches) or on pipe racks (pipe bundles on steel structure composed of one or more elevations) together with the piping.

The term "off-sites" usually refers to all these piping and cabling runs together with other non-process facilities, like utilities, storage, flare, waste treatment, etc.

AUTOMATION AND CONTROL SYSTEMS

Such systems of plants consist of instrumentation, DCS (Distributed Control System), PLC (Programmable Logic Controller), SIS (Safety Instrumentation System, formerly known as ESD (Emergency Shut Down system)), fire and gas detection systems, etc. Panels and consoles for automation and instrumentation systems including computers and monitors are usually located in control rooms. Instrument cabling with different types of connecting items and junction boxes, and instrument air tubing (when pneumatic instrumentation is used) with a variety of fittings and accessories are parts of these systems.

ELECTRICAL SYSTEMS

Electrical power supply, grounding, lightning protection, outdoor and indoor lighting, and weak current systems, including telephone, intercom, paging, badge systems, fire alarm, data, and closed-circuit television (CCTV) systems, are part of electrical facilities together with the cabling and other connecting materials for these systems. It is to be noted that some of the weak current systems like data, CCTV, and fire alarm are sometimes classified under instrumentation and automation facilities or even under separate systems such as "telecommunication" and "safety and security". The term "electrical equipment" refers to transformers, electrical generators, UPS (Uninterrupted Power Supply), HV (High Voltage)–MV (Medium Voltage)–LV (Low Voltage) panels (switchgear), MCCs (Motor Control Centers), and electrical motors.

SEWER SYSTEMS AND WASTE TREATMENT

An industrial plant has always a domestic and a storm sewer system, as made available in any other (nonindustrial) facilities, like a business center or a housing development. Domestic sewer effluents are treated in domestic waste treatment packages or discharged into municipal domestic networks. Storm water is almost always discharged back to the nature unless entrained with pollutants such as oils, chemicals, or excessive suspended solids. These are also called open drain systems. Most industrial plants additionally require oily water sewers and/or chemical sewers, the effluents of which are to be treated in proper waste treatment systems for protecting the environment in accordance with applicable laws and regulations. These are mostly closed drain systems.

SAFETY, SECURITY, AND PROTECTION SYSTEMS

These systems bear importance for the safety, security, and healthiness of plant personnel, other people living around or visiting as well as for protection of the environment and the plant itself.

Regarding safety, major systems commonly used are SIS, fire and gas detection, and alarm systems. Pressure relief systems composed of pressure relief valves, rupture disks, and blowdown valves which are often connected to flare systems, incinerators, or burn pits for possible release of combustible materials are also part of safety protection systems. Security systems usually consist of intrusion detection and alarm, CCTV, and badge systems. For environmental protection, there are various other systems in use, such as emission monitoring and leakage detection systems. Every plant possesses as a major safety protection system at least a fire protection system consisting of fire water facilities, foam systems, extinguishing inert systems, and fire brigade. This is termed as "active fire protection". Plants processing flammable gases and liquids are also passively protected by coating the surfaces of steel structures – carrying equipment and piping loads – with concrete or concrete-like proprietary material. The purpose is to delay possible collapse of the structure due to high temperatures reached during fire with the expectation that the extra time gained may be used to extinguish the fire or at least to evacuate all plant personnel.

BUILDINGS, STRUCTURES, AND ROADS

Almost all industrial plant facilities possess buildings, at least for the entrance security and administration offices. Some process units are enclosed inside buildings, like pharmaceutical units and food production units, in order to provide the required level of hygiene. In some cases, buildings are needed for protection of the operation and the operators against severe climatic conditions. Water-based utilities require to be protected against freezing in winter, otherwise necessitate significant amounts of steam or electrical heat tracing. Another example for processes inside buildings is batch processing units that require extensive labor for manual operations, including manual feeding of additives, unloading and cleaning of the entire process equipment items after each production cycle. In some cases, the production or storage dictates certain temperature and/or humidity ranges, so such plant buildings must be equipped with a proper Heating, Ventilation, and Air Conditioning (HVAC) system.

Other sections of a plant that are established inside buildings are offices, meeting rooms, cafeteria, locker rooms/showers/toilets, laboratories, substations (enclosing switchgears and partially the transformers), control rooms, MCCs, workshops, warehouses, etc.

On the other hand, for plants that have risks of hazardous (flammable, toxic, chocking, highly corrosive) vapor or liquid leakage and buildup, the processing units are always built outdoor. Examples are refineries, petrochemical and some chemical processing units. Control rooms in such plants are designed to have positive pressure inside to avoid entrance of hazardous vapors. Possible flammable vapor emissions that might enter inside would otherwise require all the hardware to comply with a more stringent area classification, leading to very high costs. Such control rooms are also designed as blast-proof.

Equipment items, pipe work, cables, and other appurtenances in industrial plants require to be supported by suitable structures. Structures mean concrete foundations, concrete or steel columns and beams, bracings, etc. to form pipe racks and equipment structures. In certain cases, equipment is installed under shelters to protect both equipment and the operation/maintenance personnel from unsuitable climatic conditions like showers and extreme sunshine.

In order to enable access of operation/maintenance personnel and maintenance equipment, like cranes, to all the needed points inside the facilities, it is obvious that a network of roads has to be constructed. Such roads are also necessary for access by emergency personnel and vehicles, like fire trucks and ambulances to emergency areas, as well as for escape of other personnel away from such incidences during emergency situations.

Regarding process plants, it is worth noting that there are no fully identical facilities even if the process is similar. This is obviously due to differences in the following:

- Process technology
- Product specifications
- Raw material compositions
- Plant capacity
- Climatic conditions
- Land characteristics and geological conditions
- Adjacent properties
- Available highways, roads, railways, ports, and pipelines
- Availability of resources including utilities and power supply
- Other geographic conditions
- Applicable laws and regulations
- Owner's standards
- Engineering standards and codes used
- Different manufacturers for equipment and materials
- Requirements of the market the products are sold

Hence, designing and constructing a plant always require special tailoring to match the conditions at hand with the requirements of the project.

HOW DOES A "PROCESS PLANT PROJECT" EMERGE?

All new projects arise from a public need, legal requirements, and/or discovery of a commercial opportunity. As the investment idea is identified, strategies and solutions are then to be developed. This usually involves making surveys, collecting data, and performing conceptual studies including review of available process technologies. Assessment of viability as well as the economics is a crucial step in deciding whether to proceed with the project or not. Major part of the above information is obtained through pre-feasibility and feasibility studies. As more data is collected and information generated, precision of the level of viability and economics is further improved. Other steps in initiating the project involves selecting the technology as well as the site, setting up project budget and schedule, obtaining management consent for the investment, and getting prepared for project execution. One of the most important challenges in having the green light for a project is obtaining financing which may also include finding partners for the investment. Obtaining financing opportunities requires having a bankable feasibility report at hand, showing that the project is viable and feasible.

Sometimes, part of the aforementioned activities is contracted out by the investors to external consultants or engineering companies, instead of doing all these in-house.

Arrangements need to be commenced for engineering, material supply, and construction stages in the form of setting up a detailed project implementation plan. Teaming up timely and adequately for project execution is always very critical for the success of projects. Even at such an early stage, there are already many activities that the investing party, i.e., the owner, must carry out, like the following:

- Looking for land (if not yet made available)
- Investigating the latest legislation requirements, having the environmental assessment study prepared
- Timely applications for local and national permits
- Arranging project financing
- Preparing the key specifications for project including performance guarantees, applicable codes and standards, terms and conditions for securing bids and contracting out services, like the engineering

A more comprehensive list for owner's activities is given in Chapter 2 – Parties Involved in Process Plant Projects.

During initiating a project, almost all of the above-outlined activities are normally carried out by project owners with or without external support of consultants or consulting firms. It is very important for project owners to set up an implementation strategy above all the detailed implementation plans at very early stages in order to define which activities will be performed in-house and how the rest will be split into different scope packages to be contracted out. There are no standard ways in doing so; however, facts like capabilities possessed by the project owners, legislative requirements, nature of the project, market availability, and qualifications of service providers, i.e., engineering contractors, consultants, vendors, and construction contractors, as well as the conditions of financing agreements, normally lead to the final decision. These conditions are further explained in Appendix 3 – Contracting Strategy.

It is also worth highlighting that owner's activities for the whole project are not limited only to the project initiation. Even if all the rest of the activities – that actually make up the very major part of the project – are contracted out, owner has to be actively present at every stage all along the entire life of the project by reviewing and approving services like engineering, equipment and material supply, and construction performed, as described in the later sections.

2 Parties Involved in Process Plant Projects

Implementation of industrial plant projects is basically a joint effort and effective collaboration by a number of parties having different expertise and different responsibilities for achieving one major end result:

- To construct a new plant or a major modification to an existing plant as intended (i.e., making the plant ready for operation on schedule, within budget and with performance targets achieved without any significant health, safety, and environmental incidents or accidents).

In other words, a project involves different entities, each having a different organization from top management to the lowest employee level with various specific functions and responsibilities for different aspects of the overall scope of the services. Each of them performs their roles for making the entity's contribution needed for constructing the industrial plant ready for operation in accordance with relevant contracts and purchase orders placed for the project.

WHO ARE THESE PARTIES?

A general listing of the main parties normally involved in industrial plant projects is as follows:

- Owner
- Licensor(s)
- Engineering contractor(s)
- Vendors
- Construction contractor(s)
- Owner's engineer or Project Management Contractor/Consultant (PMC)
- Third-Party inspection companies
- Financing organizations
- Consultants
- Insurance companies
- Permit/Approval authorities, regulators

The above list may in some cases include a purchaser and even an operator. This happens if the owner selects contracting the implementation of the entire project to another entity that will act as the purchaser for the entire services including material supply and construction. This is the case for lump-sum turnkey projects. Owner may also decide to select an entity, usually named as the operator, to operate the plant for a certain period of time under a contract. When the purchaser is contracted to proceed also with the operation of the plant, then it may be named as the Investor, as in the case of Build-Operate-Transfer contracts.

On the other hand, the above list may as well be shorter in case the owner uses its own resources instead of involving some of the above-listed parties. Owner may select such an approach when the project is simple, small, and the owner has personnel in sufficient numbers and with the required capabilities to execute the project alone.

WHAT DO THEY DO?

Brief descriptions about the functions of these parties are outlined below. Duties and responsibilities of these parties having major roles in project implementation will further be detailed in the next chapters.

OWNER

Owner is a single person or more persons in partnership or a company or a joint venture of companies that have decided to invest on a project. It is the "Client" to all other parties taking some role in the project, excluding national and local permit-granting authorities. The first kickoff for a project is made by the owner internally within its organization in order to perform the initiating functions for the investment, like the following:

- Identifying the idea that is to be worked out as a new project
- Making
  - Preliminary market surveys for raw materials and products
  - Initial selections for capacity options
  - Tentative selections for process alternatives
- Selecting a plant site suitable for the project; having surveys performed for topographical, geological, and meteorological conditions
- Contracting out or in-house rendering of conceptual or basic design that will support feasibility studies
- Investigating the applicable laws, regulations, incentives, and permits with regard to the investment, construction, and plant operation
- Having a bankable feasibility report prepared (more details are covered for Feasibility Studies in Appendix 2 – Feasibility Studies)
  - For verification and updating of the previously collected and evaluated information on market surveys as well as capacity and process options
  - For checking through total capital investment and operating cost estimates, cash flows and revenue projections to lead to estimating the Internal Rate of Return, Payback Period and Net Present Value for verifying project feasibility
  - For searching financing opportunities based on the feasibility report
- Finalizing project scope by deciding on project options, budgeted costs, and schedules
- Obtaining project funding and the approvals

- Setting up the execution methodology for engineering, procurement, construction, commissioning, and startup phases
- Setting up detailed description of the project requirements
- Preparing documentation for requesting proposals on providing the early starting services, like preparation of know-how package, engineering services, etc.
- Continuously monitoring and updating the feasibility parameters, as well as the business strategy as more information becomes available along development of the project

As the project initiation progresses, the owner needs issuing timely the RFQs (Requests for Quotations) for services in accordance with the project schedule. Meanwhile, the owner has to keep up being actively present at every stage of the entire life of the project by making reviews and giving approvals.

LICENSOR

Many manufacturing processes are covered by patents that give the holder the power to prevent others from using the invention for a number of years from the date the know-how is patented. Rights for using such manufacturing technologies are packaged with a full description of the process know-how and purchased by paying a patent rights fee and/or a royalty for the produced quantities. This is called a license, while the holder who grants exclusive rights of such use is named as a licensor. Licensed production technologies widely exist for relatively complicated processes in chemical, petrochemical, and oil-refining plants. Patented know-how is protected by law until its expiry, and it is therefore kept confidential by both the holder (licensor) and the licensee in order to get protected against unfair competition that may attempt selling or using such know-how without paying the costs of the invention. Such inventions are usually made on a laboratory or bench scale, developed in pilot plants and then improved in actual plants along the life of the patented technology. Know-how providers submit their information mostly in the form of process packages consisting of process flow diagrams, P&I diagrams, equipment and instrument data sheets, process description, instrumentation and control philosophy, specifications on operations and hazards, etc.

On the other hand, there are other types of industrial processes that are either not under or no longer under any patents. These are usually the technologies developed by the project owners in their manufacturing plants from an expired license or developed by engineering companies after execution of a number of those unlicensed or license expired projects. Such technologies are usually available for relatively simple or very common processes and often referred to as "Open Art" technologies. All of these still require having certain knowledge and experience accumulated and is kept confidential by those who own such information.

ENGINEERING CONTRACTOR

Engineering contractor (often shortly termed as the "engineer") is one of the most important actors in implementing industrial plant projects. This is due to being the party that sets out the plant configuration and all the detailed design solutions. The engineering function normally stays actively in the project from the very beginning until the last moments, up to plant takeover by the owner for commercial operation.

In some projects, scope of the engineering contractor covers only engineering or engineering plus procurement services. In such cases, the rest of the services, i.e., construction contracting, construction management, pre-commissioning, commissioning and startup, are carried out either by the owner or by other contractors of the owner.

When a job is contracted to an engineer (i.e., engineering contractor) on EPC (Engineering, Procurement, Construction) or EPCm (Engineering, Procurement, Construction management) basis, the engineer's contractual responsibilities in the project are broader than an only Engineering Services (ES) contract and often include stringent guarantees for pre-agreed plant performance criteria, together with other guarantees against material and labor defects, schedule delays, and contractual incompliances. Expiry of contractual guarantees usually happens a couple of years after plant acceptance.

Sometimes the engineer uses subcontractors for some portions of engineering works that are either specialty topics outside the field of the engineer's expertise or are particular works that are to be performed by companies possessing specific certification in accordance with either applicable regulations or owner's strong preference (like, connecting to high-pressure natural gas mains or performing specific safety studies).

For engineering services of a project, owners may sometimes decide to go with more than one engineering company. This becomes compulsory in case the project covers multiple complicated units that cannot be handled by a single engineering contractor within a reasonable period of time.

The above points are mostly valid for multidiscipline engineering companies that are organized to have almost all the necessary engineering disciplines required for engineering of a plant. However, there are cases in which the industrial plant is quite simple and sometimes, no process unit is included in the scope. For such cases, the owner may decide making multiple engineering contracts, one with each of single discipline engineering companies in civil, mechanical, and electrical engineering. Such an approach likely reduces cost of the engineering as such companies are rather small with relatively low overhead costs, but usually at the expense of certain level of weaknesses in other practices, such as effective coordination and integration of engineering among those different disciplines concerned. Collaboration of a number of different companies with different working standards and at different office locations often creates significant drawbacks in quality and schedule expectations.

Not all engineering furnished for industrial plants is rendered by engineering contractors. Proprietary designs for utility packages, waste treatment packages, security systems, etc., and many items of equipment such as heaters, compressors, pumps, turbines, etc., are engineered and designed by manufacturers (vendors) who are experts in such fields, and are continuously developing their products through R&D and collected feedbacks from their units under operation. For such items, engineering contractor only specifies the necessary operating and design data together with type and required performance of the unit, so that vendors can carry out detailed design and fabrication by using their own proprietary methods. Engineer's duty is then limited to the review of the design and committed performance including verification of inputs, outputs, materials of construction, scope, manufacturing methods, etc., in order to eliminate any potential omissions or mistakes against the contract requirements. Similarly, shop drawings for steel structures, vessels, rebar schedules for concrete, etc., are prepared either by vendors or construction contractors depending on the split of the contractual scope in the project.

Figure 2.1 shows the major stakeholders taking part in a project mainly in regard to the interfaces with the owner/PMC as one important epicenter and EPC/EPCm contractor as the other.

VENDORS

The name "vendor" refers to suppliers or manufacturers of systems, equipment, and materials. For purchasing project items, usually the engineering contractor prepares the project vendor list at very early stages of the project and establishes a mutual agreement with the owner on these listed companies to which bid requests will be issued. The list covers all categories of materials and equipment items as well as system packages that are required for the project. The list normally contains several vendor names for each separate category unless there are any specific reasons to go to a single nominated vendor. The latter situation may arise when there is alliance

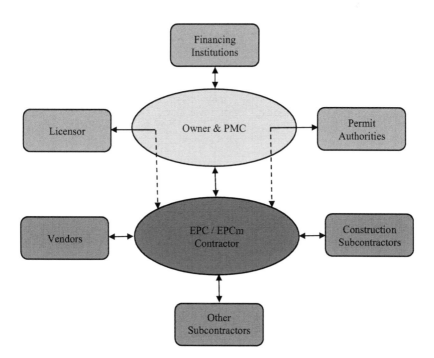

FIGURE 2.1 Major stakeholders taking part in a process plant project.

agreement between the owner and such vendor, or the owner has preference for that specific vendor due to certain reasons, like previous good experience with that vendor or for minimizing the spare parts inventories by having already installed the same brand of equipment elsewhere in the plant. Nominating a vendor is almost inevitable when a new complicated system, like Distributed Control System (DCS), is to be connected to the existing main system of the plant.

Good collaboration by each of the vendors is one of the critical points for the success of a project, not only for getting the desired performance and quality, but also for timely and properly completion of the engineering that requires receiving well-prepared and complete vendor drawings from which certain data is to be incorporated in the engineer's drawings.

Construction Contractors

These are the companies that carry out the construction works of the project which generally cover the following:

- Site preparation (excavation, landfill, piling, etc.)
- Underground (U/G) works (sewers, U/G piping and cables, electrical grounding grids, etc.)
- Civil works (construction of foundations, structures, buildings, roads, fencing, etc.)
- Mechanical works (installation of equipment, piping, package units, steel structures, heating, ventilation, and air conditioning, plumbing, fire protection, etc.)
- Instrumentation works (installation of field instruments, DCS, programmable logic controller, safety instrumentation system, security and data systems, instrument tubing, cables, cable trays, conduits, junction boxes, etc.)
- Electrical works (installation of transformers, generators, motor control centers, switchgear, uninterrupted power supply, grounding and lightning protection, lighting, weak current systems, cables, cable trays, conduits, etc.)
- Insulation and painting

Construction works of a plant are either entrusted to a single contractor or split into a number of specialty packages, like civil works, mechanical works, electrical and instrument works, etc., in accordance with the construction strategy set up during early stages of the project.

Construction is one of the last coming, as well as a very important, phases of a process plant project since all the extensive time and efforts spent up to that stage including engineering and procurement are merely for constructing the plant.

Owner's Engineer or PMC

Owner may select hiring a person or persons to perform part of its duties in a project for and on its behalf, with reporting responsibilities back to the owner. Employment of owner's engineer mostly happens on cases when the owner does not possess sufficient manpower with the required qualifications. Main duties often include a sort of technical due diligence along the project to fill any gaps in expertise within the owner's project organization.

A deeper approach in this respect can be to assign a PMC for delegating notable part of owner's functions in the project. A PMC is an engineering company, other than EPC contractors, specialized in process plant projects. PMC's functions usually cover reviewing of and commenting on engineering, procurement, and construction activities with reference to contractual requirements and liabilities of the vendors and contractors, basically pertaining to scope, quality, health-safety -environment, progress, schedule, and budget. PMC either gives approvals – with or without comments – or makes rejections through review of activities and/or deliverables of the contractors. Employment of a PMC mostly happens for large complicated projects. PMC performs its duties under a detailed contract with the owner and fundamentally, manages EPC contractors of the project.

Sometimes, it is the financing organization of the project that obliges owner to have such a third party employed in the project. In such a case, the position is also named as "lender's engineer."

Third-Party Inspection Companies

All purchase orders placed to vendors and contracts awarded to construction companies include detailed specifications and references to codes for inspection and testing in order to achieve the intended quality. With such an objective, inspection companies are often employed by the owner or the engineer to control the quality of the supplied material, manufactured items and the construction. Such employment is in reality an additional measure to enforce vendors' and contractors' contractual responsibilities for performing the specified inspections and tests. Third parties may also be employed by vendors and construction contractors as their subcontractors for carrying out quality control requirements on their behalf.

Financing Organizations

It is not usual to have process plant investments executed on 100% shareholders' equity, mainly due to the fact that capital expenditures (capex) for process plant projects are quite high. "Debt/equity" ratios usually vary from 60/40% to 85/15%. Accordingly, for putting in the project a portion of the total investment cost, project owners look for low-cost financing prospects. These are usually provided by national and international banks as well as export credit agencies of most Western countries. A very important prerequisite set forth by financing organizations for granting such a loan or other credit facility is to have a bankable feasibility report showing the project is remarkably feasible. Noticeably, there are other important factors, like financial standing of the investor, political stability of the investor's country as well as the country where the plant site is located. Some lenders, upon granting an investment loan, may appoint engineers for monitoring the design, construction, and commissioning of the works, titled as "lender's engineer," as described above.

Please refer to Appendix 5 – Financing, for more details on financing schemes.

CONSULTANTS

In general, consultants are persons or a group of persons who give professional expert advice to people on business, law, technology, etc. In process plant projects, consultants may be employed by owners to take advice on various matters including market surveys, feasibility analysis, contract preparation, technology selection, cost estimating, applicable regulations, legal issues, etc. The other parties involved in a project may similarly need to use consultants for getting advices on matters out of their expertise. Due to the nature of the knowledge and experience possessed, engineering contractors may be employed as consultants on most project matters.

INSURANCE COMPANIES

Several different insurance policies have normally to be purchased by the parties taking part in the project. The most commonly required policies are the following:

- Professional liability
- Cargo insurance
- Workmen's compensation
- Construction all risk and erection all risk
- Third-party liability
- Machine breakage

The purpose of buying these policies is to minimize negative impacts of possible adverse incidences/accidents on the success of the project. In such situations, insurance companies are involved in estimating damages, if any; negotiating claims and reimbursing the insured parties for those damages.

Please refer to Appendix 4 – Contractual Clauses, for deeper discussion.

PERMIT/APPROVAL AUTHORITIES

A number of permits are required during implementation of an industrial plant project. The type, nature, and application requirements for those permits obviously depend on the country and the region where the plant is constructed. Such permits are usually for the following:

- Site selection and production facility establishment
- Environmental impact assessment
- Construction
- Emissions
- Discharge to natural waters (sea/lake/river) and soil
- Water intake (from sea/lake/river)
- Fire brigade
- Trial run
- Plant operation
- Other (natural gas allocation, electrical power allocation, etc.)

Thus, the site location, the process, the engineering, construction materials and construction itself should strictly comply with the applicable laws, regulations, and directives. Permit application documents are mostly quite sizable dossiers that cover information about the project, including process inputs, process outputs, the process itself, emissions and wastes, certain critical engineering calculations and drawings, most of which are relevant to health, safety, and environment. These are reviewed by relevant authorities or regulators and audited before granting permits.

Another important matter that owners apply for approval is for "investment incentives." The availability, nature, and extend of incentives usually depend on the type and size of investment and the region the investment will be made. Incentives are normally in the form of reduction of taxes, like custom duties, value-added tax and deferring some others, like corporate taxes.

3 Contracts

A contract is a written agreement between two parties specifying liabilities and responsibilities in a deal of undertaking by one of the parties the supply of services, technology, facilities, material, or construction for the other party. A contract is sometimes named as an agreement.

WHAT IS IT IN THIS BUSINESS?

During implementation of industrial plant projects, owners require to sign contracts with other parties for services, material, technology, financing, or construction. The name of such an agreement sometimes changes to Purchase Order (PO), especially for the supply of materials (including equipment) and services. Both, though have a similar objective and approach in terms of specifying liabilities and responsibilities, the latter differs in being mostly unilaterally signed and issued by the purchaser. Both are prepared often with reference to a technically and commercially conditioned, negotiated and accepted bid. A PO requires receiving vendor's confirmation of the order acceptance in writing. Furthermore, material and services ordered through a PO have usually very specific item by item-defined limits for scope and the content. The end product is then packed up (service products arranged in dossiers) and transferred to the purchaser. Ordered items are checked at certain stages at vendors' premises via inspection and expediting visits by the purchaser (i.e., the owner) or its delegates until the items are received by the purchaser, mostly as fully completed supplies. On the other hand, contracts are widely used for mostly longer lasting works that require day-to-day relations of the contracting parties along the contract period. The scope details are usually shown on drawings and sometimes also in Bill of Quantities (BOQ) which may change along the progress of the project. Owner has to continuously review and comment on the developing output of a contract, at least the key deliverables, and so, extensively contributes in shaping it up as required.

HOW IS IT AWARDED?

A contract is mostly based on an RFQ (Request for Quotation) or an ITB (Invitation to Bid) by the owner that is quoted by owner selected bidders, then conditioned by the owner (or by the owner's consultant) and finally, negotiated with the successful bidder to draw up the contract. It is important to note that normally the successful bidder should not necessarily be the one proposing the lowest overall price, but it is the one who:

- Really understood the project requirements.
- Has sincere interest to do the job.

- Possesses available resources required for the project.
- Has previous experience and success on similar jobs.
- Is well established in all internal operations with proper policies, proce-dures, and standards.
- Possesses necessary certifications and good records, e.g., on quality assur-ance/quality control and health-safety environment.
- Proposes a good organization scheme that looks to work to the satisfaction of the job requirements.
- Submits a reasonable time schedule.
- Is financially fit.
- Does not give a "claim orientated" impression.
- Estimates the size of the wok correctly (e.g., total man-hours and its breakdown).
- Covers the entire scope with reasonable prices.
- Does not make unreasonable/unacceptable qualifications to (deviations from) the contract terms.

More details about RFQs, bid conditioning, negotiations, and contracts are covered in Chapter 11 – Construction Contracting. So, now let us move on to briefly mention-ing about major contract types.

WHAT ARE THE MOST COMMON BASES?

Contracts are based either on competitive bidding or on direct negotiation with a selected or nominated bidder. The former approach is more commonly followed and apparently requires that there is more than one bidder and all are in more or less similar size and possess similar qualifications. For selecting to go with direct nego-tiation, there must be sound justifications, like fully satisfying previous experience, only a single acceptable bidder available to quote, obligation to purchase from a nominated bidder imposed by the financing, etc. A way in-between may be the "right of first refusal" granted to a preferred bidder. In this approach, the preferred bidder is asked to comply with the price and conditions acceptable to the owner, while the bidder knows that owner may as well proceed with competitive bidding if a mutual agreement is not smoothly reached.

Contracts are most commonly on one of the following pricing bases:

- Lump sum (LS)
- Lump-sum turn-key (LSTK)
- Unit price (UP)
- Combination of the above
- Other (like cost plus)

"Lump sum" basis means the price for the entire services and/or material and/or construction is fixed and firm from the beginning of the contract up to the end. Such a basis can be used in cases where the scope, nature, and extent of the work are very well defined. A basic advantage to the owner is the motivation imposed on contractors for finishing the work as early as possible so that payments are received earlier, while savings on certain indirect and overhead expenses can be achieved by minimizing any possible schedule delays. Another advantage is increased expectation by the owner to have the cost better controlled. In most cases, such LS/LSTK contracts also include unit rates (like unit prices for man-hours, construction equipment, material, etc.) to be used for possible scope changes to be added on top of the fixed price if such happens so.

The basis is named as "EPC" or "EPC LSTK" (Engineering-Procurement-Construction Lump-Sum Turn-Key) when a contractor undertakes full responsibility of doing the entire work including engineering, procurement, materials and equipment supply, fabrication, construction, erection and often also commissioning for which all costs – along with a profit component – are to be paid under the contractual LSTK price and contractual payment schedule.

"Unit price" basis means the price for each unit quantity of services, material, and construction is fixed, while the quantity depends what is actually consumed to complete the scope of the contractual work. Of course, there is always an estimate of quantities required in such contracts and the contractor needs to demonstrate to owner's satisfaction the actual quantities consumed during the execution. The actual quantities used are measured during the project execution and are multiplied by corresponding contractually fixed unit prices to form the total price. This basis is often used when the work cannot be described in sufficient detail for a LS/LSTK pricing at the time the contract is awarded. An advantage of such basis is to have much less cumbersome handling of contractor's extra work requests and claims by the owner or its representative who in such cases should, however, check very carefully the actual quantities of work done. Another advantage is to enable the owner to start a contract earlier than possible, compared to going along LS or LSTK basis. It should be noted that the risk of having project delays increases for this type of a contract, even though there is always an agreed contractual schedule. This is mostly due to the fact that contractors may tend moving their resources from less time sensitive projects to other projects that are deemed to be more critical for them in terms of impacts of possible delays.

Combination of LS/LSTK and UP is selected when certain parts of the project are clearly defined to enable LS/LSTK pricing, while some other parts are not.

Additionally, there are various other types of contracts used in process plant projects. These are mainly the following:

- Cost plus fee
- Cost plus fixed fee
- Cost plus fixed fee with bonus malus schemes
- Guaranteed maximum price

The "cost plus" type of contracts are based on paying contractor actual documented costs plus a fee for general overhead and profit. This type of contracts is sometimes used at early phases of projects, like basic engineering and is then converted to LS, considering that sufficient information for converting the basis to LS (or LSTK) pricing has eventually been achieved so that the contractor can, at such point, prepare a proper and justifiable LS/LSTK bid.

For more details on contract types and contents, please refer to Appendix 3 – Contracting Strategy and Appendix 4 – Contractual Clauses.

CONTRACTING STRATEGY

WHAT IS IT?

A possible definition of contracting strategy is "a high-level plan to define how to meet the requirements expressed in a requisition or in a complex initiative." It shows how the project will be implemented with the characteristics of the contracts that will encompass it. It determines the success of a project because it takes into account the risks and the best way to mitigate them.

WHAT POINTS SHOULD IT ADDRESS?

In order to be successful, a good contracting strategy should address the main strategic elements of the project implementation, such as the following:

- Breakdown in elements of work (work breakdown structure – "WBS," work packages)
- Number and localization of engineering – design office
- Extent of RFQ definition
- Early engineering
- Early procurement for long lead items
- Guidelines for contracts
- Definition of contract type

WHAT SHOULD IT BE BASED UPON?

The contracting strategy is not an abstract exercise; on the contrary, it must come from an accurate analysis of the project, its risks, opportunities, constraints, as well as past experience in similar projects. A partial list of such points is as follows:

- Safety
- Time/cost balance
- Availability of resources in number and skill
- Local conditions and regulations
- Financing arrangements
- Execution in a joint venture

FORMALIZATION

In order to be effective, the contracting strategy should be well thought of and should be the result of an interactive dialogue between the person/function in charge of its definition (usually the project manager) and various functions that will have to support him/her with various specialist inputs. Usually, this requires a trial-and-error type of process and should culminate in a written document, which will form part of the set of project documents according to the applicable quality assurance manual.

Very importantly, the requirement for the contracting strategy to be formalized in an approved document is not an expression of bureaucracy, but rather a system to ensure that the strategy will be well thought of, shared, so maximizing its chances to be effective.

The above are the main concepts about contracting strategy. For a more complete analysis, including a discussion of the type of contracts and their field of application, please refer to Appendix 3 – Contracting Strategy.

CONTRACT ADMINISTRATION

Contract administration is a relatively young discipline in the project management world.

The singular aspect of this discipline is that it includes activities that have always been carried out in executing a project. The innovation lays with the fact that while in the past they were done by "somebody," in recent times the business trend is to concentrate them in the hands of a dedicated function. These areas are grouped together because of their affinity to one another, so creating a new professional profile, which is in charge of the administrative and contractual aspects of the management of a project.

From a functional standpoint, there is no difference between this and other disciplines, like project control, or engineering or others. All of them have the responsibility to execute specific portions of the project, reporting to the project manager. The latter is accountable for everything, and is actually delegating some of its operating responsibilities to specific components of the project team. To visualize this concept better, it is sufficient to think of a small project, where the project manager covers various areas, often including project control and contract administration, or even – in some cases – construction. This is clearly dictated by economic considerations, to save on the project cost, specifically on the staff functions. As soon as the size of the projects becomes larger, it can no longer be managed with the "one man show" concept, and it is necessary to use more sophisticated schemes of responsibility delegation.

The main lines along which the contract administration acts are as follows:

- Administrative aspects
- Contract management
- Change management
- Claim management
- Business functions

For a more detailed discussion of contract administration, including its position in the corporate organization of a company and its professional aspects, please refer to Chapter 18 – Contract Administration.

4 Management

Management plays a crucial role in all types of profit and nonprofit business activities. It can simply be defined as the act of getting the necessary people together to accomplish desired goals in line with certain principles by using resources, such as materials, tools/hardware/software for management and business and, of course money, as needed.

It is also regarded as the following:

- The art of making people more effective than they would have been without it
- The science of how to do this

The word "manage" is stated to be inherited from the Italian word "maneggiare," meaning "to handle," which in turn was derived from the Latin word "manus" meaning "hand."

Management has vital importance as one of the major ingredients in implementing industrial plant projects. All the parties that contribute to the accomplishment of a project have different levels of management, such as the following:

- Senior management (sometimes referred to as top management or upper management)
- Middle management (reporting to senior management while supervising lower management)
- Line management or lower management (such as department, section, or team leaders)

Effectiveness of these levels is a very important parameter for the success of a project.

MANAGEMENT OF AN ENTERPRISE

Each enterprise has an organizational structure with various management levels and positions set up according to the actual requirements specific to that company. Despite having a number of different definitions made by experts, management functions are widely accepted to consist of the following:

- Planning
- Organizing
- Staffing
- Leading or directing
- Controlling an enterprise

From a theoretical point of view, classifying the management functions as above may look all right. However, these, in practice, have significant overlapping portions that cannot be totally separated from one another, with each function affecting the others.

Management set up in an enterprise varies extensively from company to company, but often comprises the following:

- Operations
- Marketing and selling
- Purchasing
- Finance and accounting
- Planning
- Staffing, motivating, and servicing
- Information technology
- Auditing and legal counseling

The configuration of the management in an organization is set up and implemented in accordance with the mission, vision, objectives, and policies of the enterprise.

Under each of the above branches, there are usually several sub-branches, like engineering often reporting to operations. In most enterprises, these main branches are directed by general managers, managing directors or presidents who are, in turn, reporting to the Chief Executive Officer (CEO) and board of directors assigned by shareholders.

It should be kept in mind that an enterprise may possess the most modern operating tools, a highly skilled and experienced staff, well-prepared policies and vast financial resources, but may still fail in making successful business. There may be many reasons and circumstances leading to the failure; however, the mostly encountered cause is poor management.

Management is apparently a very vast topic on which there are extensive studies made by universities, private institutes, and management experts. Further details on this topic apparently remain outside the scope of this book, excluding a very specific and critical type for execution of projects. This is "project management" that is recognized as a distinct discipline due to the specific nature of project execution requirements.

PROJECT MANAGEMENT

Project manager is, no doubt, the lead actor/actress in implementing projects. Although there are lately quite a number of graduate programs available in many universities and a variety of education programs given by private institutes on this discipline, he or she is actually trained on the job. This is especially inevitable for complicated projects, such as the ones for industrial/process plants. For this type of projects, project managers mostly start their career in some other, but usually a related engineering position, like project engineering, until getting to the necessary level of maturation in terms of knowledge and experience.

Project management is a specific type of management that has in time been developed for successful execution of projects. As described in the earlier paragraphs of the book, execution of projects is quite different from ordinary non-project business operations. Unlike the latter type of business activities, projects have a transient nature with a certain life time until arriving at the completion. In other words, projects are undertaken for a specific purpose with a defined beginning and an end. None of them are identical to any previous, current, or future project. Such differences require development of distinct skills and adoption of a separate management approach.

More detailed description of project management function is made in Chapter 6 – Project Management.

Part II

Launching the Project

5 Starting the Implementation

In Part 1, a short review was made on some general aspects of process plant projects. Now, let us start taking a closer look into it.

For describing how projects are implemented, there are basically two main perspectives that require paying simultaneous attention as follows:

- One of them is to view stage by stage, from major project phases, like engineering, procurement, construction, etc., to main group of activities, like disciplinary tasks of various functions such as engineering and procurement, and then, zooming into further details for each of those subitems.
- The other perspective is looking out for the project activities and developments that normally take place along execution in a chronological order, by keeping focused on how a project most likely goes, from the very beginning all the way through, up to the end.

This book aims combining both of these perspectives, with the target of creating simultaneously the sense of time and the key project execution features accomplished in implementing process plant projects.

PROJECT INITIATION

Project initiation is the first step in life of all projects. In general, it starts with a real current or an expected future need and sometimes just from an inspiring idea that may lead to shaping up to a project. For the case of process plants, the starting point of the project initiation is usually based on educated guessing and thinking about what would be the best to produce for the next couple of decades. Ideas coming out are further developed through literature, process and market surveys, preliminary calculations and sometimes, even bench scale and pilot plant experiments. The main objective is verifying the economic outcome through a feasibility report which is based on the information that could be made available until then. A feasible looking idea together with the desire to invest will keep the project moving to subsequent steps following the owner's start with the project initiation as mentioned earlier in Chapter 1 – Preamble and Chapter 2 – Parties Involved in Process Plant Projects.

Almost all of the above-outlined activities are normally carried out by the investing party (i.e., the project owner) with or without external support by others. This stage represents the real and the first initiation step of the overall project, while it should be noted that any other party joining in the project at different times along the project progress has its own project initiation, comprising the preparation and organization of activities for the collaboration.

Engineering companies join a project at early stages, after being contracted by the owners for a specific role, like as follows:

- PMC (Project Management Consultant/Contractor)
  Such an assignment is common for large size, complex investments, sometimes as an obligation from financing conditions of the project.
- Engineering (if scope is limited to only engineering services)
  - Basic engineering or FEED (Front-End Engineering and Design)
  - Detailed engineering
  Both types of engineering services are normally required in a process plant project.
- Engineering with extended scope responsibilities (if scope also covers subsequent project activities such as procurement and construction management or construction)
  - EPCm, or
  - EPC
  Owner is to evaluate and decide which of the two approaches suits to the project best.

Project initiation activities of engineering companies are basically composed of the following:

- Assignment of a project manager
- Contract review at the beginning of the contractual services
- Teaming up for the project
- Making up project strategies and contract budget
- Start preparing detailed project schedules
- Holding internal kickoff meeting
- Arranging project coordination procedure
- Preparation of basic engineering and design data
- Setting up the filing system
- Holding kickoff meeting with the owner
- Preparation of project execution plan

In an engineering company, a project manager (PM) is appointed by the relevant divisional director for each contract. Immediately after his or her assignment, the PM starts the project initiation of the contract. One of the first tasks is to go over available project information, starting by reviewing the contract unless he or she has already been involved in the proposal and contract award stages. A contract review is usually made between the PM and the proposal manager for highlighting critical as well as unclear points. A simultaneous activity is to start building the project organization, giving priority to the functions that will shortly begin producing services for the project (like the engineering disciplines, project control, and procurement). Detailing of the project plan is also immediately started, especially with focus on home office services at the project initiation.

A principal milestone in project initiation is the kickoff meeting with the owner. It is important for the engineering contractor to get fully prepared for that meeting. The preparation includes compiling and internally distributing all the relevant project information available, holding the internal kickoff meeting and having project team members take out questions and options to be discussed during the kickoff meeting with the owner. Meanwhile, available information and project requirements are summarized in a formal document, usually titled as "Basic Engineering and Design Data" for reviewing and completing during the same meeting. PM of the engineering contractor prepares for owner's approval the coordination procedure specifying contact persons, formats to be used, reporting needs, etc. to be applied during the project execution. This is a simple, but very useful, document in organizing and letting everybody know from the beginning how to handle such everyday matters along the progress of the project. Another important activity by the PM at the initiation step is the preparation of the "project execution (or implementation) plan." More information about this plan is covered in Chapter 6 – Project Management.

PROJECT PHASES

Implementation of process plant projects goes through a number of phases which are named in various ways depending on the way of categorization. As a generic description for project management, a project is stated to be phased as follows:

1. Project initiation
2. Planning
3. Execution
4. Monitoring and control
5. Completion

It should be noted that stages 2, 3, and 4 above are looped to cycle many times along the life of the project until arriving at completion.

Another way of defining project phases may be based on the epicenter of the main execution work and so, becomes the following:

- Home office services
- Site services

The steps above start in the indicated order but do not necessarily occur as "finish to start" type. In most projects, home office services covering engineering, procurement, constructions contracting, etc. continue for some time after the construction site is opened.

From another perspective that focuses the characteristics of the work, the phasing can be defined as follows:

- Basic engineering or FEED including process package
- Detailed engineering
- Procurement

- Construction contracting (or subcontracting)
- Construction including pre-commissioning
- Training of plant personnel
- Commissioning and start-up

Similar to the first two classifications above, the latter phases start mostly in the given order and proceed with certain overlaps (or with extended continuation). For instance, detailed engineering, procurement, and construction contracting extends often to early stages of construction, while training of plant personnel is often performed during later stages of construction. One clear exception is the first and second stages as detailed engineering is almost never commenced before completion of the prior step, the basic engineering, because the latter is normally built upon completed, approved and so, frozen basic engineering or FEED.

PROJECT STRATEGY

Every project is different from another with respect to various aspects, like plant type, size, scope, location, particular requirements, etc. Moreover, there is one more factor that makes a project quite different from another similar project. This is the "strategy" set up for the project execution. Strategies are basically major plans or decisions made up at early stages for specifying main parameters on how to run the projects. Most crucial parameters are relevant to the overall project plan, forming high strategic levels, like deciding to implement the entire job in-house or through some contracts. There are also lower-level strategies, mostly relevant to each particular phase of the implementation like, maximizing the purchase of material from a particular country for complying with the conditions of financing.

The differences brought in projects by applying different strategies should normally affect only the implementation methodology, not the plant when it is completed. However, these are such decisions of high importance that affect the success factors, like the cost, time, quality, and even the end of the story (e.g., whether the project gets to completion or not). There is no single or just a few standard strategies to be followed; selections are to be made from quite a number of options closely related to the characteristics of the project. The strategic decisions relevant to the overall approach should come as early as possible in the beginning of a project, obviously to be revisited and modified as required along implementation based on the latest conditions and forecasts.

When an investor (owner) is ready to take forward steps in implementing a project, the parameters that are considered with utmost care are budget, time, quality, and resources. All these critical project parameters are closely interrelated and require to be well established usually through extensive reviews and internal discussions much earlier in the project. Sound strategies that are believed to be the best fit for the success of the project are to be devised before the project starts progressing. It should be noted that the parties other than the owner do also need to properly set up their own project strategies for their roles in the project as all these different parties are literally the members of the overall project organization that should run for the single main target, the project success.

Scope of a process plant project may cover only quite minor modifications to an existing plant. On the contrary, it may as well encompass a grass roots investment, such as a chemical complex or a refinery. Apparently, implementing the former type of a project is relatively easier, less time taking and does not necessarily need elaborate procedures and sophisticated strategies. Those projects are often executed by the owner's team, usually without any appreciable support from external parties. However, for the latter case, it is necessary to set up effective and detailed strategies on how to deal with all main activities of the project. In other words, as the size and complexity of the project increase, it becomes more critical for the project to make up sound and sufficiently detailed strategies.

As Engineering, Procurement, Construction (EPC) services of a process plant project normally cover the following:

- Project management
- Engineering and drafting
- Planning and cost control
- Procurement
- Health-safety-environment (HSE)
- Quality management (QM)
- Construction contracting
- Construction and pre-commissioning
- Construction management
- Commissioning and start-up

It is necessary for project owner to make best fit strategies on how to have all those services performed as successfully as possible. For medium to large size projects having notably complicated process technologies, it becomes necessary for the owner to contract out almost all those services, one way or another. Main contracting schemes for such projects usually have first a PMC assigned to the project for managing subsequently selected contractors to work on one of the following basis after a FEED is made ready by either the PMC or a FEED contractor

- EPC, or
- EPCm

Or less likely,

- EP (Engineering and Procurement Services) or
- ES (Engineering Services)

For relatively narrower contracting schemes, like the last two items above, the rest of the work is either performed by the owner or separately contracted out to specialized contractors. It is worth mentioning that the owner has to assign from its organization, regardless of the contracting scheme applied, a PM and sufficient number of specialists for following up the performance of the contractors by giving comments and approvals or rejections to the contractors.

A PMC contracting is almost always utilized when the project is quite big in size, e.g., made up of multiple units with significant complexity, like a new refinery or a chemical/petrochemical complex. The PMC manages the work for and on behalf of the owner by starting with preparation of or contracting out the conceptual design or directly the FEED and then, contracting out the EPC packages. Though the owner is still involved, PMC approach helps decreasing the size of the owner's project organization.

In certain other cases where the size and complexity of the project permits, owners may select proceeding directly with EPC contracts and do all contractor management activities by themselves.

One of the crucial points in starting implementation of a project is the process technology to be used. For relatively simple and non-patented (open art) technologies, this is either provided by the owner, based on its other plants under operation, or the engineering contractors that have sufficient knowledge and experience on that specific technology. For patented process know-hows, the technology is purchased from a licensor. As project owners obviously require ensuring that targeted feasibility of the plant is achieved, technology selection is to be carefully carried out for avoiding use of not yet proven process know-hows or the ones that are becoming obsolete.

PLANNING

In the course of implementing industrial plant projects, the strategies cannot alone lead to arriving at the desired destination. The actions to be taken along all phases of a project require detailed planning. In fact, planning is a road map of progress for projects. It helps clarifying, focusing, and forecasting a project's development by also highlighting possible pitfalls as well as hidden opportunities. A comprehensive and realistic plan will not though guarantee success; however, lack of such plans will almost always lead to failure. At the start of project activities, a project execution plan based on project requirements is to be prepared by each contractor and submitted to the owner for approval. Preparations of this kind of comprehensive plans are one of the critical duties of the PMs and represent the most important baselines of the project's quality plan.

Planning is not the only duty of the PM. He or she is though responsible for ensuring that all planning requirements of the project are correctly addressed, all other disciplines involved in the project are obliged to have their own part of the work properly planned. It is the planner, usually assigned from the project control or planning discipline, who prepares those necessary planning schedules by receiving inputs and comments from all the disciplines involved.

When a plan is approved by the owner, it becomes the benchmark for monitoring and controlling the actual progress of the project. The actual progress with respect to the approved plan is periodically verified and reported for devising proper measures corresponding to any possible shortfalls that might have developed. An approved plan may never be modified unless a contractual approval from the owner is obtained based on accepted justifications, like scope changes or occurrence of force majeure. Further information is presented on project planning in Chapter 9 – Project Control.

Project plans are prepared in a variety of forms, some of which are narrative and others are either graphical or tabular. Narrative plans, like project execution plan, HSE plan, etc., describe the targeted developments by indicating where, how, when, with whom, with what, how much, and why. Graphical formats include S-curves (progress curves), bar charts, logic networks, etc., prepared as a function of time and sketched from the information presented as tabulated calculations shown in relevant tables.

UNIT FOR QUANTITY OF SERVICES

Time spent for project services need to be quantified for planning, monitoring, and control purposes and in some contracts, directly for invoicing. The unit commonly used for such needs is the "man-hour."

A man-hour is the total time of "1 hour" spent by one or more direct persons (e.g., half an hour by each of two persons) on a particular project activity. The term "direct" is used to differentiate "persons who are producing services directly for the project" from the others who are classified as general overhead, like company upper management, general accounting, human resources, general information technology services, general secretarial services, etc. Direct man-hours are recorded for project accountings, while indirect ones contribute to the project economics in the form of overhead cost.

A project has always an estimated number of man-hours for each specific category of services, like engineering and procurement. The man-hour budget based on the estimating is a benchmark used for continuously checking the actual man-hour expenditures and forecasts with the budgeted figures all along the project up to completion. For proper control, it is a common practice to have detailed man-hour breakdowns for each discipline taking part in the project. The man-hour unit is also commonly used in construction and commissioning works of projects for direct labor and supervision, as split into each different trade of work. In addition to being a parameter for monitoring and control of the progress and cost, man-hour estimates are also utilized for resource leveling.

In some cases, particularly for construction related services, it may be deemed more practical to use larger units, like "man-days" or "man-months." One man-day corresponds to a certain number of man-hours depending on the standard daily working hours in the project. If daily working time in the project is 10 hours, then one man-day is equal to ten man-hours. Similarly, one man-month corresponds to the number of man-hours worked in a month, typically 170–200 for home office and 220–250 for jobsite.

JOB ACQUISITION PROCESS

From the description of the above processes, it is clear that executing a project is a very time- and resource-consuming activity. Many of these resources are internal to the company, and need to be kept busy working on projects and generating revenues for the company. Missing this, the cost associated with such resources (salaries, social contributions, indirect costs) would negatively impact the company's result, and if not properly balanced by revenues, they may quickly drag the company's financials into the red.

Therefore, it is essential that the company equips itself with a structure that takes care of "feeding the machine" with a continuous flow of new orders, to make sure that resource availability is matched by resource consumption. This is a very delicate function in the company, which is working in very close co-ordination with the senior management, as well as with corporate planning, to constantly update the resource availability – new orders balance.

The typical product of this continuous evaluation is the workload projection, a graph similar to the one shown in Figure 5.1.

The three areas of the graph represent, respectively, are as follows:

- The man-hours consumed by the projects currently under execution
- The man-hours associated with those projects which are currently in a tendering phase, but the acquisition of which is considered very probable
- The man-hours associated with those projects which are currently considered as potential, with lower probability of acquisition

The horizontal line corresponds to the current baseline of man-hours available.

The interpretation of the graph shows that the current projects will produce a declining workload until their natural conclusion. The probable projects would (or hopefully "will") add workload for another period of time. The situation of the company looks reasonably good, as the workload will only start to go below, say, 80% in a year or so. This outlook defines the company's appetite for new projects and is utilized by the management to determine how strong the commercial effort should be.

The job acquisition process is under the responsibility of sales people and proposal people, usually both reporting to a commercial director. For more details on the processes, the commercial considerations and the professional profiles, please refer to Chapter 17 – Job Acquisition Process.

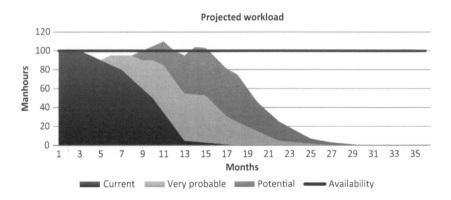

FIGURE 5.1 Workload projection.

6 Project Management

In project-type operations, project manager (PM) is the principal leader of all other leaders in the project team. He or she is responsible for all activities of the entire project and drives all phases of the job by his or her specific project management knowledge, skills, and experience. In this respect, project management is regarded to be a mission for achieving successful completion of a project. A PM is assigned at the very beginning of a project and stays on active duty until the project is closed out. In this respect, the PM can be regarded as the most important actor/actress in project implementation.

BRIEF HISTORY ABOUT PROJECT MANAGEMENT

It is predicted that project management has been practiced since early civilization. It had been generally on civil engineering and defense projects until the industrial revolution which practically started within the second half of 18th century. Scientific management theories by Frederick Winslow Taylor followed by planning and control techniques by Henry Gantt (widely known from Gantt charts) and predictions of management functions set forth by Henry Fayol have been the rising path of project management. Later, in 20th century, more elaborated and systematic tools have started to be applied due to arising needs by more complicated projects of the time. These include development of the Critical Path Method, Program Evaluation and Review Technique, and advancement of cost engineering. Meanwhile, international associations and institutes for project management were founded to support the development of project management profession.

ASSIGNMENT OF THE PM

For process plant projects, almost all of the entities taking part in the project, like engineering contractors, owners, construction contractors, etc., establish a kind of project management organization. This type of organization requires, first of all, assignment of a PM immediately upon a contract award and then, allocation of other staff from specialty functions, arranged in a matrix scheme.

The matrix scheme is composed of columns and rows, where each column represents every single project in a multi-project running environment. The rows represent specialty functions and the disciplines that produce services for the projects. In an engineering company that has normally several ongoing projects at a time, it will be seen that each of those projects have an assigned PM, who may be assisted by a Project Engineering Coordinator (PEC) or at least one or more project engineers (PEs). The PM of one of the projects, say project no. 1 managed by PM1, represents one of the columns of the matrix together with the assigned PEC or PE while other columns are headed by PMs of the other projects. On the rows, there are the

TABLE 6.1
Organization Matrix of Project Management

	PM_1	PM_2	......	PM_i	PM_{n-1}	PM_n
F_1	√	√	√	x	x	√
F_2	x	√	x	√	√	x
......	√	√	√	√	√	√
F_j	√	√	√	x	√	√
......	x	√	√	√	x	√
F_m	x	√	√	√	x	√

specialty functions (such as engineering, procurement, project control, construction) and disciplines of these functions (like process engineering, piping design, planning, inspection-expediting, etc.) that have certain scope of work in the project. This is shown in Table 6.1.

Please note that the above figure shows projects from 1 to n, with functions and disciplines from 1 to m. The check sign, "√" and the cross sign, "x" stands for "yes" and "no," respectively, to indicate whether a function has any scope in each of those projects.

Such a matrix works in the way that each PM looks after and undertakes full responsibility for usually a single project while the functional departments have to organize themselves for serving to all the current projects, the scopes of which require their department's collaboration. Therefore, the head of each such discipline needs to assign a lead engineer or a manager from his/her department for ensuring sufficient care is taken for each project that the department has to work on.

ORGANIZATION

A PM is assigned from the pool of PMs by the relevant divisional director when a new project starts. The matrix shown in Table 6.1 is re-sketched in a simplified organization chart form in Figure 6.1.

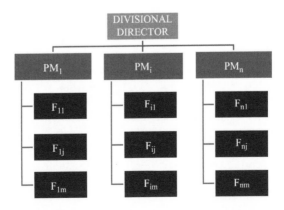

FIGURE 6.1   PM assignments for projects.

As execution of every project is to be specifically tailored to suit the particular needs of each, project organization is also set up in accordance with those needs of the project. Thus, every project does not require to be staffed similarly with regard to participation of functions and specialty disciplines.

As stated above, project organizations are established by assignments from all the necessary disciplines and functions required to be in the project. Based on the size and particular needs of the project, these key positions are filled in sometimes by lead engineers or sometimes coordinators or managers to whom a number of other staff assigned to the project reports. It is to be noted however that, the personnel assigned from the disciplines to the project are not permanent subordinates of the PM. Though the PM is the head of the team for the entire life of the project, the project organization disappears when the project ends. In fact, all the disciplines and functions have their own managers who technically follow up and support, as needed, their personnel working in project teams with regard to technical and administrative matters.

A simple organization chart showing the main reporting lines in an Engineering, Procurement, Construction (EPC) process plant project can be sketched as shown in Figure 6.2.

In order to keep the above figure simple, staffing details below the functional groups (i.e., boxes at lower levels for coordinators, engineers, designers, buyers, supervisors, etc.) are not shown.

As project execution methodologies have continuously been advancing in parallel to the increasing expectations of the investors, project organizations are set up to cope with the new requirements. Accordingly, further specialized positions are brought in project team configuration in line with the latest working practices of the contractors as well as the characteristics of the projects. Figure 6.3 shows a further staffed scheme compared to Figure 6.2. Again, the staffing details below the major functions are omitted for clarity of the chart. The positions in Figure 6.3 are briefly described in the following.

PROJECT MANAGER

Reports to higher management of his/her company, such as a divisional director. Roles and responsibilities are described in this chapter.

QUALITY ASSURANCE/QUALITY CONTROL

Responsible for the quality assurance (QA) and quality control (QC) of the project. This position is lately regarded to be out of the normal project matrix due to concerns

FIGURE 6.2   A simple project organization chart for EPC services.

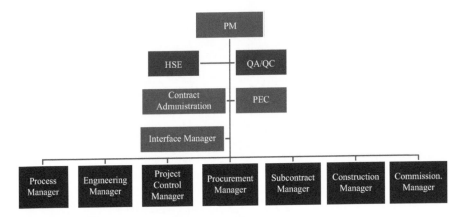

FIGURE 6.3 A typical project organization chart for EPC services in medium to large process plant projects.

for keeping neutrality and independence of the QA/QC against other project execution objectives that may compete. He or she reports to quality assurance manager of the company while assisting the PM, like a consultant and an auditor in the project on quality matters by making continuous follow-ups, audits, and reporting.

HEALTH, SAFETY, AND ENVIRONMENT

Responsible for the health, safety, and environment (HSE) matters of the project. This position is also regarded lately to be out of the project matrix due to the same reason mentioned above for the QA/QC position. He or she reports to HSE management of the company while assisting the PM like a consultant and an auditor in the project. Similar to the situation for QA/QC, some companies do not impose any reporting lines of HSE to the PM due to same concerns outlined above.

PROCESS MANAGER

In some engineering companies, process engineering is considered as a totally different engineering function, as separate from the other engineering disciplines and indicated in organization charts, accordingly. As described in Chapter 8 – Engineering, process engineering generates the basics of process plant configurations in the beginning of projects while rest of engineering groups mostly produce more of a standard kind of detailed design, structured on the process configuration devised by the process engineering.

PROJECT ENGINEERING COORDINATOR (PEC)

Responsible for coordination of home office project work, basically the engineering and drafting activities of the project. PEC usually acts as an assistant to the PM. When a PE assigned to a construction site, the position name becomes field project engineering. Please see Chapter 8 – Engineering.

CONTRACT ADMINISTRATION

Employed often in organizations for large and complicated projects. The corresponding functions of this position have been handled until lately by other members of the project team and the same approach is still applied in small to medium projects. Further details about this function are presented in Chapter 18 – Contract Administration.

INTERFACE MANAGER

Such position is especially needed for projects having multiple execution centers, i.e., various home office locations and/or multiple construction areas with a number of different entities taking part in projects in the same plant. Those entities are usually joint venture partners or subcontractors of the main EPC/EPCm contractor and/or other contractors of the owner having interfaces within the plant.

ENGINEERING MANAGER

This position manages all the specialty engineering groups. Lead specialty engineers of engineering departments (e.g., mechanical, project, piping, instrumentation, civil, electrical), design drafting coordinator and document management staff report to this position. If such a position is not present when the project is simple and small, the coordination function of the foregoing positions usually passes to the PEC. The engineering function and the engineering disciplines are described in detail in Chapter 8 – Engineering.

PROJECT CONTROL MANAGER

Basically, responsible for planning and cost control activities of the project and may have planners, cost engineers, and estimators assisting this position in the project. Further information is provided in Chapter 9 – Project Control.

PROCUREMENT MANAGER

The function deals with purchasing, vendor expediting, shop inspection, and logistics for materials and equipment. The coordinator or manager for this function is assisted by buyers, expediters, inspectors, and logistics clerks. Detailed description of this function is presented in Chapter 9 – Procurement.

SUBCONTRACTS MANAGER

This function carries out contracting services mainly for site works, such as fabrication and construction work. If required, other services, like engineering and drafting are also contracted out by this function. In relatively small projects, this position may be handled by procurement as the techniques used in both functions are quite similar. Further information is presented in Chapter 11 – Construction Contracting.

CONSTRUCTION MANAGER

During the home office services of the project, the position is usually named as home office construction and deals with activities, like constructability reviews until construction site is opened. Then, a resident construction manager is introduced in the project organization together with a number of field supervisors and supporting engineers. Further information is available in Chapter 12 – Continuation to the Project at Field and Chapter 13 – Progress of Project at Field.

COMMISSIONING MANAGER

Commissioning joins the project usually just before pre-commissioning and mechanical completion activities. More information is available in Chapter 13 – Progress of Project at Field.

DUTIES AND RESPONSIBILITIES OF THE PM

Basic duties and responsibilities of PMs can be outlined briefly as follows:

- Planning
- Organizing
- Monitoring
- Control of major project parameters, like the following:
  - Cost
  - Time
  - Resources
  - Scope and
  - Quality

for completing the project with all contractual requirements fulfilled on time and within budget, hence, accomplishing a successful job that satisfies the owner. This generic description is shown in Figure 6.4.

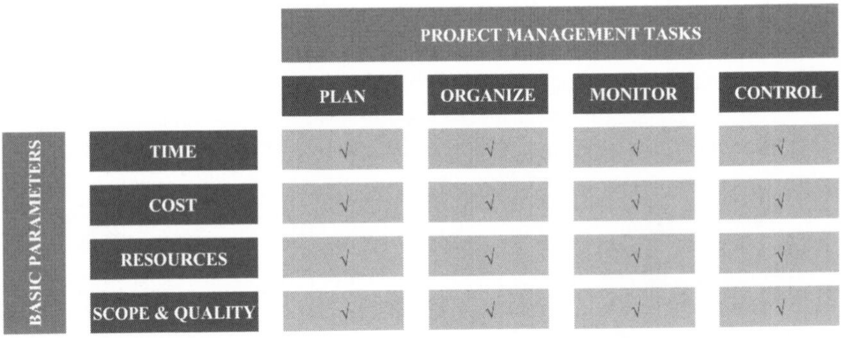

BASIC PARAMETERS	PROJECT MANAGEMENT TASKS			
	PLAN	ORGANIZE	MONITOR	CONTROL
TIME	√	√	√	√
COST	√	√	√	√
RESOURCES	√	√	√	√
SCOPE & QUALITY	√	√	√	√

FIGURE 6.4 Project management main tasks with respect to basic project parameters.

As already outlined in Chapter 5 – Starting the Implementation, the basic tasks may also be expressed as follows:

- Project initiation
- Planning
- Execution
- Monitoring and control
- Completion

The above tasks, excluding the first and the last, are to be repeated as many times as needed until completion.

The duties and responsibilities of PMs shown above are a little generic and usually referred to as the classical or theoretical definition. Below is a more detailed job description for an engineering contractor's PM on duty in executing process plant projects:

PROJECT INITIATION

- Make contract review and handover meeting with the proposal manager before starting the project.
- Start setting up the project organization in collaboration with the disciplines that will participate in the project.
- Make up project strategies; distribute the budget to all the disciplines to take part in the project.
- Distribute to the project team the available project information including the relevant parts of the contract, existing data and drawings, owner's specifications, etc.
- Have project control start preparing project schedules and cost follow-up studies.
- Organize an internal kickoff meeting with the disciplines.
- Have coordination procedure and basic engineering data prepared by PEs.
- Have the filing system set up.
- Contact owner's representatives and organize the project kickoff meeting for the project start.
- Lead the project kickoff and subsequent project meetings with the owner and have the minutes prepared.
- Prepare the PEP (Project Execution Plan).
- Have drawing lists, requisition index and job specification list prepared by design-drafting coordinator and PEs.
- Start having the periodical project reports prepared by PEs in collaboration with project control.
- Closely follow up the timely start of the process engineering and initial preparations of the other engineering activities.
- Review, comment, and approve all project reports, schedules, minutes of meetings, and other project documents produced.

PROJECT DEVELOPMENT—HOME OFFICE SERVICES

- Continue leading the project by making all contractual correspondence, arranging and participating meetings with other parties involved in the project.
- Continue managing all project execution activities in line with approved schedules, budgets, and other contractual requirements.
- Ensure that engineering activities are performed in accordance with applicable specifications with timely flow of the required information among the disciplines involved.
- Initiate and follow up any necessary remedial actions, when required, for putting back the project on schedule, within budget and free of omissions / mistakes.
- Ensure owner's comment and approvals are timely received for the engineering deliverables issued.
- Follow up timely performance of the engineering reviews, e.g., P&ID, plot plan, model, Hazard and Operability (HAZOP), constructability, etc., as well as preparation of material requisitions, construction requisitions, and technical evaluation of the bids.
- Review and approve vendor lists and then, bid tabs prepared by the purchasing group after receipt and evaluation of the bids, receive owner's approval if contractually required.
- Ensure inquiries and purchase orders are timely and correctly issued and expediting-inspection of materials and equipment are being performed in line with the contractual requirements.
- Review and approve construction contractors (subcontract bidders) lists as well as bid tabs prepared after receipt and evaluation of the bids by the contracting/subcontracting group; receive owner's approval for the same if contractually required.
- Ensure construction subcontracts are timely awarded.
- Submit for approval and negotiate with the owner all contract changes;
- Review change requests and claims received from vendors and subcontractors.
- Review and authorize invoices to the owner; give final approvals to vendor and subcontractor invoices.
- Approve and issue progress reports; prepare presentations for project reviews.

PROJECT DEVELOPMENT—SERVICES AT CONSTRUCTION SITE

- Arrange timely moving of the project to the jobsite.
- Have construction management organization headed by a resident construction manager assigned for managing construction and subcontractors; mobilize the organization to site.
- Follow up construction of temporary facilities at the jobsite.
- Organize and attend construction kickoff meetings with the construction subcontractors.

- Follow up for timely receipt of the engineering deliverables and construction materials at the jobsite and take necessary remedial action if required.
- Review with the resident construction manager the progress of site works as well as any issues on quality, HSE, and other areas of concern for taking corrective actions if needed.
- Review with the resident construction manager and the contract administrator the contractors' extra work requests, claims, disputes, back charges, etc.; make decisions on approvals or rejections of these.
- Follow up the performance of the resident construction manager and review the same for the rest of the site team with the resident construction manager.
- Attend management review meetings at site with contractors and the owner.
- Continue managing the project and making the contractual correspondence with all stakeholders involved.
- Move to site during critical times, such as plant turnarounds and reported performance disputes of any of the parties at site.
- Continue managing the project execution in line with the approved schedules, budgets, and other project requirements.
- Ensure the construction is properly progressing to mechanical completion with all necessary pre-commissioning activities being performed.
- Check that contractual training requirements are timely addressed.
- Make sure necessary arrangements for the forthcoming commissioning and startup phases are made in accordance with the contract.
- Submit for approval and negotiate with the owner all contract changes.
- Review change requests and claims by subcontractors.
- Review and authorize transmittal of invoices to the owner; follow up the review and approval of contractors' invoices at site.
- Issue progress reports including the site work.
- Follow up achievement of mechanical completion milestone and have the mechanical completion certificate signed by the owner.
- Have as-build drawings completed and handed over to the owner.
- Start demobilizing construction subcontractors and the construction management team.
- Follow up execution of commissioning, startup, and performance testing. Make sure that commissioning, startup, and performance testing activities are performed successfully on schedule, within budget, and in compliance with all other contractual requirements.

Project Closure

- Demobilize the remaining construction management team and the commissioning team.
- Close construction contractors' contracts and commence the contractual guarantee periods.
- Make necessary arrangements for handing over the plant to the owner.
- Ensure all administrative aspects of the contract as well as the subcontracts are completed (payments, retentions, back charges, liquidated damages, etc.).

- Prepare the contract close out report.
- Hand over all contractual documentation and certificates and close out the contract with the owner.
- Have plant handover/provisional acceptance certificate signed by the owner.

HOW DOES THE TEAM WORK?

During home office services of an engineering contractor, the members of the team from various disciplines normally work in their own department areas for the projects. Alternately, all team members including the PM may be organized as a "task force" to move to a specific location in the office, called "task force area." The latter working style is usually beneficial in achieving a better control over the work during its progress. Controllability is more critical when there are outstanding project constraints, like tight schedules and/or budgets as well as particularly complicated scopes and/or highly confidential process technologies. Consequently, a sufficient number of specialists with required qualifications is assigned to the task force, though it is unlikely possible to feed up all task force members continuously with work at all times. On the other hand, the advantage of the former option is the flexibility of using available resources more efficiently for the company by assigning those specialists to more than one project when possible and so, reducing idle time as well as creating options for giving support to critical projects when needed.

As noted earlier, implementation of process plant projects is an integration of several different fields of specialty knowledge and experience that require extensive and continuous flow of information among all disciplines involved during the entire life of the project. Especially, the engineering and design is quite critical in that respect. No specialists or specialty groups can complete their part alone without continuously receiving from and giving to other disciplines project information and comments generated as the design progresses. For instance, the structural group requires at least preliminary piping (scope) layout in order to start designing a piping structure inside a unit. When the first issue of structural design is available, this is distributed to several other disciplines. Electrical and instrumentation specialists need using the structural design during their cable tray routing. The same information is also fed back to piping for re-checking any mismatches against the latest status of the piping design. All engineering disciplines receiving design information, often in form of updates, make reviews, and give back comments about any inconsistencies that either exist or may likely arise along progress of design due to process modifications, incorporation of vendor drawings, results of stress analysis, etc. Equipment engineering disciplines also need to follow up design of both piping and structural disciplines, especially for nozzle orientations and auxiliaries that are external to equipment, like platforms and ladders. These latter details are also of interest to electrical and instrumentation specialists for routing and locating their hardware. Hence, cycles of design updates and corresponding comments continuously flow within the team of specialists for shaping up to the final design.

TEAM BUILDING

One of the important success factors in implementing projects is the level of team work that can jointly be created by the participants of the project. A process plant project is one of the best examples for such jobs in which integration, collaboration, and coordination should be maintained at very high levels all along the project. This means the following:

- Exchange of information on project objectives, plans, developments, and critical issues are effectively done.
- All participants get well familiarized with each other and maintain effective lines of communication.
- Teams with a thorough knowledge of the objectives and close working relations with each other perform much better in terms of time and quality, and so the project economy.

According to many owners and engineering contractors, team building sessions are fundamental to success and so are organized as dedicated meetings for information sharing and critical issue debating by the key project members at certain stages of the project. In addition to team building meetings, recreational events, like leisure trips or dinner programs may as well be arranged for the project participants including owner's project staff. These are all for having everybody in the project get along well and improve collaboration.

PROJECT EXECUTION PLAN

The project execution plan (PEP) is a detailed road map for the project, prepared in the beginning to describe basic project information, contractual requirements and major plans for coping with challenges of the project. It covers all the project phases by way of making systematic reviews and forecasts up to project completion and targets identifying possible critical issues for taking sound precautions at early stages of the project. In this respect, it is one of the major supplements of the project's quality assurance and represents a quality assurance plan for the project. The PEP should normally cover the following main topics:

General

- A general presentation of the project and the scope
- Policies on Quality Management and HSE
- Special project constraints, secrecy agreements, insurance requirements, etc.
- Surveys and similar specific activities/studies already made
- Criteria for plant acceptance
- Basis of estimate
- Change management

ENGINEERING

- Basic engineering data
- Scope of process engineering and design
- Scope of other engineering disciplines
- Design drafting
- Subcontracting plans and scope, if any
- Key nominations in engineering and design drafting
- Applicable codes, standards, and regulations
- Applicable engineering documentation already available, including owner's specifications
- Engineering reviews to be made
- Critical engineering issues and corresponding precautions planned
- Engineering quality assurance

PROCUREMENT

- Scope of procurement
- Critical procurement issues, critical supplies, and corresponding precautions planned
- Key nominations in procurement
- Procurement strategy
- Qualification requirements for vendors
- Vendor (bidder) list
- Procedures for purchasing, expediting, vendor inspection, and logistics
- Vendor material and test certification requirements
- Spare parts, lubricants, and catalysts
- Material handling and supply acceptance
- Procurement quality assurance

CONSTRUCTION CONTRACTING (SUBCONTRACTING)

- Scope of construction contracting
- Construction contracting strategy
- Qualification/certification requirements for contractors
- Contractors (bidders) list
- Key nominations
- Critical contracting issues, critical activities contracted, and precautions planned
- Contracting quality assurance

CONSTRUCTION

- Scope of construction
- Construction management
- Subcontracting strategy
- Construction assignment plan and key nominations

- Critical construction work
- Constructability
- Field facilities and temporary facility requirements
- Warehousing
- Material inventory control
- Permit to work system
- Heavy lifting
- Construction quality assurance plan
- Material and test certification
- Punch listing
- Mechanical and provisional acceptance
- Plant handover plan

COMMISSIONING AND STARTUP

- Commissioning scope
- Commissioning strategy
- Key nominations
- Acceptance criteria
- Plant turnover
- Training

HEALTH, SAFETY, AND ENVIRONMENT

- Critical HSE issues of the project
- HSE during home office services
- HSE throughout construction
- HSE during commissioning and startup
- Security of personnel, information and facilities

PROJECT SCHEDULES

- Look ahead schedules
- Overall project schedule
- Progress schedules

PROJECT ORGANIZATION CHARTS

- Home office organization
- Construction organization
- Commissioning and startup organization
- Overall project management team organization

COORDINATION PROCEDURE

Coordination procedure is described in the next paragraph.

Overview and Conclusions on Critical Issues and Safeguards for Success

(This is the final summary or abstract of all critical issues and corresponding measures considered.)

PEP is a concise collection of facts and forecasts and is to be distributed to all project team members as well as to the owner in order to make it serve as a protocol for common understanding about the project. After it is issued, it may undergo a couple of revisions due to either important project developments or incorporation of owner's comments.

COORDINATION PROCEDURE

As mentioned in Chapter 5 – Starting the Implementation, coordination procedure is prepared by the PM of the engineering contractor or his/her delegate (PEC or PE) in the beginning of the project for specifying contact persons, project formats, reporting requirements, etc., to be applied all along the project execution. This is an essential document for organizing and letting everybody know from the beginning how to handle everyday matters during working on the project. The procedure should typically cover the following:

- Official name of the project to be used throughout the project documentation
- Names, project positions, phone numbers, and e-mail addresses of key project staff of both the owner and the engineering contractor
- Contract number(s) assigned to the project
- Official project language
- Guidelines for communication and preparation of the minutes
- Document distribution matrices
- Document numbering system
- Approval requirements for documentation produced
- Filing
- Reporting
- Invoicing
- Authorized signatures
- Main forms and formats to be used throughout the project

Coordination procedure generally refers to project implementation matters between the owner and the engineering contractor by focusing mainly the home office part of the services. The reason is that the organization and key names are often not yet exactly determined for the construction and the commissioning stages. Anyway, it is regarded as a must to also prepare a site coordination procedure with each site contractor for the construction and commissioning stages.

For further information on project management, the reader may refer to Project Management Institute's publications at https://www.pmi.org.

Part III

Developing the Project

7 Project Development during Home Office Services

A process plant project is rendered for the main objective of establishing an efficient, operable, reliable, maintainable, safe, environmental friendly and so, the desired production plant. It also requires spending the best efforts to keep it on schedule and within budget. In order to construct such a plant, it is essential, during the project development, to:

- Engineer, design, and specify the material (bulk materials, equipment, packages, other plant systems, and components) for supplying these through requisitions, drawings, specifications, and material take-offs.
- Carry out a systematic procurement campaign for purchasing technically acceptable materials at the best prices, with acceptable terms and conditions, and within acceptable delivery times, including the supply of vendor information for completing the engineering.
- Engineer, design, and specify the construction in sufficient detail for carrying out a systematic construction contracting campaign to facilitate selection of capable construction contractors at justified prices.
- Engineer, design, and specify the construction in the form of drawings (layouts, plans/sections/details, schematic diagrams, etc.) and specifications in necessary details to enable construction contractors to correctly and efficiently do the construction work in line with the schedule and within budget.

The above activities are the "engineering," "procurement," and "construction contracting" services, almost always carried out at the home office and usually by engineering contractors, unless the project is decided to be implemented differently, e.g., by the owner alone and/or directly at the jobsite.

The next paragraphs outline the progressive development of these activities. More detailed descriptions are given in the next chapters which focus on each of the different disciplines taking part in these activities.

PROCEEDING WITH THE ENGINEERING

Engineering involves mostly complex clusters of orderly and interrelated activities, needed for specifying the entire plant configuration in detail and for facilitating material supply, construction, and finally, the operation of the plant. Accordingly, it covers engineering deliverables, either on paper or in electronic file form, consisting of calculations, design (drawings, data sheets, etc.), and specifications including requisitions and material take-offs in order to effectively guide the physically more concrete products of the entire work, comprising the following:

- Material to be supplied
- Construction to be accomplished
- Startup of the plant

Thus, engineering is without a doubt the crucially important component of industrial plant projects though it corresponds to a small fraction of the total investment cost. Without properly rendered engineering, it is unlikely to avoid ending up with any kind of serious problems, like extensively time taking and expensive re-doings or even sometimes impossible to totally rectify situations that drastically hinder the plant operation. Such adverse consequences may include the following:

- Very difficult to resolve clashes
- Improperly performing equipment and systems
- Health, safety, and environment (HSE) risks
- Incorrect layouts causing cumbersome operation and maintenance
- Low capacities, yields, reliabilities, availabilities, and/or efficiencies

Development of a project starts usually after preliminary engineering during the initiation stage. The upfront work that forms such initial engineering, as mentioned in Chapter 5 – Starting the Implementation, usually covers the following:

- Process selection
- Preliminary/conceptual design, basically for feasibility studies, investment consents, incentive and/or financing applications, investment permits, etc.
- Other critical studies, like risk assessments, environmental assessments
- Process package to be purchased from the licenser (if the process is under a patent)
- Review of the process package and engineering checks for the license information

The above is followed by basic engineering or front-end engineering and design, and then detailed engineering. Procurement and construction contracting activities are started later as engineering information necessary to proceed with these are prepared and released.

A typical progress curve for a process plant project is shown in Figure 7.1. The initial activities correspond to the starting portion (the left-hand-side part) and the final activities are represented by the completion portion (the right-hand-side part) of the progress curve, which is also known as the S-curve.

The above graph is only indicative as the exact form of the S-curve changes from project to project, sometimes moving with a much lower or higher slope at certain portions of the S. This means, the curve may extend either for a shorter or longer period of time and with different rising trends at the beginning as well as in the

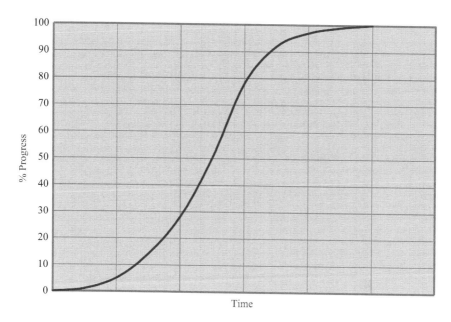

FIGURE 7.1 The project progress curve (also named as the S-curve).

middle and at the last portions, corresponding to initiation, development, and completion of the project, respectively. In any case, project progress at the initiation and the completion sections of the curve is normally much slower than the development section in the middle. This is due to various reasons that include having much fewer people working on the project during these two stages compared to the time frame when many disciplines are extensively producing services during the development stage. Additionally, early periods require certain time for learning, investigating, and getting acquainted with the job while completion requires making the final small efforts or touch ups on last pending matters with few participants who are capable to carry out such finalization.

The project development stage corresponds to the faster rising section in the middle of the project progress curve. In reality, there is no specific milestone generally agreed for a solid discrimination between project initiation and project development; however, this presumably happens a while after the project kick off meeting when fundamental design decisions are made and process engineering commences. After such a milestone, engineering departments become ready to start working one after another in coordination with each other, with all the available project information received, discussions made on how to proceed, and the essential basic plant configuration is practically set up.

For process plant projects, project development commences practically with the issue of piping and instrumentation diagrams (P&IDs) and the process data sheets for equipment and instrumentation. As soon as the process engineers issue the P&IDs as the process discipline's issue, mechanization of these diagrams are started. In many organizations, it is usually the project engineers who mechanize the P&IDs, which is simply rearranging the contents to make them more suitable for detailed engineering. Mechanization activities are followed by line numbering, line listing, and equipment list preparation. At this time, the instrument discipline can commence marking up instrumentation and automation comments on the P&IDs together with tagging and listing instruments. Again, after the first P&ID activities, piping can start with the scope layout preparation that will lead later to the preparation of piping layouts and further details, either on two-dimensional computer-aided design (2D CAD) or through three-dimensional (3D) modeling. These project developments can be started not only in piping design, but also in all engineering disciplines.

Distribution of process equipment data sheets enables mechanical discipline engineers to start further specifications of such equipment and preparation of equipment requisitions to be used in procurement. Vessel engineers work on static equipment and specify drums, tanks, columns, etc., while rotating equipment engineers work on specifying pumps, compressors, fans, turbines, etc., and heat transfer engineers design heat transfer equipment which are mostly shell and tube heat exchangers, double-pipe heat exchangers, air coolers, etc.

During this time, the structural and civil engineers commence designing and calculating the structures that do not necessarily require receiving any vendor information. For plants inside buildings or plants having service buildings, it is necessary to have first the architectural layouts and sections/views of these buildings prepared as well as loads of plant components estimated in order to proceed with the static calculations.

The actual start of electrical engineering usually happens to be a little later than most other disciplines since it requires certain prior information generated by other disciplines, like layouts and power consumption estimates. So, electrical engineers in the early stages often make a rather slow start by preparing general specifications, preliminary layouts, single-line diagrams based on estimated electrical loads, etc.

Almost all the above activities require prior definition of standards, codes, and job specifications as well as the applicable regulations to be used in the project. Some of those, especially the ones for structures and HSE requirements are dictated by the laws and regulations applicable in the region where the plant is to be constructed. However, some, such as application to the project American, European, or other standards in piping, equipment, and instrumentation, are usually at the option of the owner. It is important to note that some owners and all well-established engineering companies have their own engineering standards, practices, and/or guides for engineering and design.

Materials to be used in the project are usually specified by material engineers in relation to compatibility to the process conditions including the nature of fluids and/or solids involved and marked up on flow sheets that are then called MOC (Material of Construction) drawings. These drawings become the reference in specifying all materials for piping, field instrumentation, equipment, supports, other bulk material, etc.

The engineering deliverables are registered in logs, named drawing list, piping isometric list, requisition list, and job specification list. The lists concerning drawings are commonly handled by design-drafting coordinators and the rest by project engineers, usually through a document management group that records, keeps track of the revisions, and distributes all drawings, including the ones issued by vendors. The number of drawings in a small to medium size project may reach to hundreds or sometimes a few thousand. This figure may go up to some ten thousands for very large projects.

The team of people carrying out the engineering activities is composed of engineers, designers, and drafters working for the following departments:

- Process engineering
- Project engineering
- Instrument and automation (control) engineering
- Piping engineering
- Material engineering
- Vessel engineering
- Heat transfer engineering
- Machinery (rotating equipment) engineering
- Architectural, civil, and structural engineering
- Electrical engineering

Engineers do the calculations, make main engineering, and design decisions, selections, and reviews including technical bid reviews and vendor drawing reviews. Designers are experienced technicians preparing mostly the layouts, selecting design details, making take-offs, preparing bulk material requisitions, and solving design detail problems while drafters are CAD operators, drafting the design in accordance with the design-drafting standards used in the project.

The activities mentioned so far are only part of what normally happens during the engineering development. More detailed information about this stage and job descriptions of engineering disciplines are given in the next chapter. At this point, let us repeat for the process plant projects that the engineering is an integrated performance of various different engineering disciplines. Each engineering discipline transmits and receives certain information for developing their part, during which re-checking and revising the previous design as required are practically regular aspects of this type of work. This cycle of checking and revising frequently happens in projects due to the usual project developments, like receipt of vendor information after order placements, identification of clashes or mistakes/omissions, achieving the actual data that was previously estimated, change requests made by owner, etc.

PROCUREMENT

Procurement is a function that starts during early stages of project development and goes hand in hand with the engineering progress. As the engineering completes the specifications of equipment and bulk materials for quotation by the bidders (manufacturers and suppliers), purchasing activities can be started with the issue of the inquiry packages.

An inquiry package usually consists of the following documents:

- Request for quotation
- Instructions to bidders
- General purchasing conditions
- Requisition for the items and attached data, specifications, and drawings

Request for quotation and instruction to bidders advises bidders the bidding procedure including the details for the requested format of the bid as well as the bid closing date. General purchasing conditions generally set forth rights and liabilities of both the seller and the purchaser in the case of the placement of an order. The requisition (sometimes called the material requisition) is the technical part of the entire package that describes the requirements about the supply, including the technical documentation to be produced by the vendor. The documentation, named "certified vendor information," represents a crucial input for the detail engineering, which cannot be progressed without incorporating the actual vendor data of the ordered items.

Before commencing the preparation of inquiries, there are a couple of important steps to be taken. The first one is setting up and receiving the management's/ owner's agreement on the procurement strategy and the procedure to be applied in the project. The strategy is the set of major options selected for material supply activities, like where and on what basis to buy. The procedure details each stage of procurement specifying how it will be carried out, followed up, approved, and reported. The next step is the preparation of the bidder list (usually called the "vendor list") for each inquiry. This is usually done jointly with the owner (as owner's approval is almost always a contractual requirement) and is based on a list of pre-qualified manufacturers/suppliers.

On the bid closing date, the bids received are opened by a committee of representatives, usually from the purchasing disciplines of the engineer and often the owner (unless the project is on lump-sum turn-key basis). The bid information is recorded as received and signed off by the committee members before unpriced bid copies are sent to relevant engineering disciplines for technical evaluation. Hence, the bid evaluation is made along two different paths, one being the technical conditioning while the other is the commercial conditioning. During the conditioning process, the objective is to clarify ambiguous or missing points, to correct bid mistakes and adjust all the bids to the originally intended basis set up in the inquiry package. The next major step is placing the order to the successful bidder whose final standing is determined to be the best.

The above outlined activities are performed by the purchasing group of the procurement discipline. There are usually three other subgroups inside the procurement department, namely the expediting, inspection, and logistics. These groups participate in the project after order placement by the purchasing group. The functions of the procurement discipline are covered in detail in Chapter 10 – Procurement.

CONSTRUCTION CONTRACTING/SUBCONTRACTING

During progress of the engineering, field works should normally be started at a certain progress level of the project that allows starting the site works, without waiting so long for all drawings to be issued for construction. Such an overlapping approach, when planned properly, serves to shorten the overall project completion time in line with the so-called fast track program. Depending on the construction strategy, construction scope of a project may be split into a number of separate contracts, instead of selecting a single construction contractor. A multi-contracting scheme may cover separate work packages, like site-grading, civil works, fine works for buildings, mechanical works (including equipment, piping, and structural steel works), instrumentation, and electrical works. In some cases, construction work is split into more packages compared to the above, such as equipment erection, piping fabrication and erection, structural steel fabrication and erection, insulation, and painting.

Construction contracting follows techniques which are very similar to the ones used in purchasing. Before commencing the preparation and issue of inquiries, it is necessary to set up and agree with the owner on a procedure to be applied in construction contracting. Similar to the procurement procedure, this document details how each stage of contracting activities will be carried out, followed up, approved, and reported. Then the bidder list is to be prepared for each inquiry. The list is normally based on a list of pre-qualified contractors, similar to the case of procurement. For bidders whose available qualification information dates back more than a couple of years, it is advisable to make re-qualification before including such potential contractors in the bidders list. If construction contracting is managed by an engineering contractor working on an Engineering-Procurement-Construction (EPC) or Engineering-Procurement-Construction management (EPCm) basis for the project, then the term becomes construction subcontracting, and the selected construction company is called a subcontractor.

A construction inquiry is usually made up of the following documents:

- Invitation to Bid (ITB) or Request for Quotation (RFQ) or Invitation to Tender (ITT)
- Instructions to bidders
- General contract conditions
- Special contract conditions
- Construction requisition
- Attached drawings and specifications

Invitation to bidder is a cover letter presenting the work and referring to any previous communication in which the bidder confirmed interest to quote. Instruction to bidders specifies the bidding procedure, the requested format of the bid, and informs bidders of the general content of the bid package. It further specifies bid closing date and pre-bid requirements like the site visit. General contract conditions set forth rights and liabilities of the parties involved in case of a possible order. Special contract conditions indicate in detail all the special construction requirements at the project site, including temporary facilities, HSE, pricing techniques for extra works, etc. The construction requisition is the technical part of the entire package describing the technical characteristics of the material supply, fabrication, and construction by making reference to all relevant drawings and other technical documentation enclosed as attachments.

Further details about construction contracting are given in Chapter 11 – Construction Contracting.

OTHER HOME OFFICE SERVICES

There are other disciplines that participate in the project implementation at the home office. These are mainly the following:

- Project control
- HSE
- Quality management
- Home office construction

Below is a brief outline about these disciplines for the purpose of completing the general picture on the project development phase at the home office.

Project Control

This function deals with scheduling and cost engineering all along the life of the project. Services of the same discipline are also needed during proposal preparation.

Scheduling is the follow-up of the baseline schedule of the project initially set up when starting the work. Accordingly, the progress is monitored and incorporated in the schedule for comparing actual status with the plan. Cost engineering involves continues follow-up of expenditures and updating forecast for making evaluations on cost trends as well as verifying the latest forecasts against the latest budget figures.

The reports prepared by project control are almost like the dashboard of a car in demonstrating continuously the status of the performance and the speed of moving forward. Further details are covered in Chapter 9 – Project Control.

Health, Safety, and Environment

Health, safety, and environment have long been a matter of attention along the industrialization of the societies and have been under increasing level of care by laws and regulations. Within the last decades, more detailed methodologies and regulations are set up for ensuring elimination of the risk exposures to human health, job safety, and environment.

Risk identification and evaluation are made for the process technology, engineering, design, construction, commissioning, startup and operation of the process plant. More specifically, P&ID and layout reviews, HAZID (Hazard Identification) techniques, HAZOP (Hazard and Operability) reviews, and SIL (Safety Integrity Level) analysis are the most common techniques used in the identification, evaluation, and safeguarding against potential hazards. The main HSE responsibility for engineering is borne by line management and subordinates of all the engineering disciplines taking part in a project with supporting consultancy and auditing by HSE specialists. More details are given in Chapter 15 – Health, Safety, and Environment.

Quality Management

Like HSE, quality assurance and quality control responsibilities also lie with the line management of all the disciplines taking part in the project execution while facilitated and audited by quality management staff. The objective of the quality management is basically to ensure that the products/services are designed, supplied, and constructed to meet or exceed the requirements. Quality management discipline performs the following functions to fulfill this objective:

- Provide support to the management activities of the project from quality system point of view
- Provide support to production of project services from quality system point of view
- Perform quality audits; analyze the operations and products, like drawings, reports, and data
- Verify compliance of all project activities with the quality system requirements (e.g., ISO 9001)
- Follow-up on NCR (Non-Conformity Reports)

More information is presented in Chapter 16 – Quality Management.

HOME OFFICE CONSTRUCTION

Construction discipline starts functioning at the home office before the construction site is opened. Such early duties of this specialty service mainly cover supporting the following disciplines:

- Project management—in setting up the construction strategy and supporting project execution in construction matters
- Engineering—in carrying out constructability reviews and comments
- Project control—in preparation of construction schedules
- Construction contracting—in bidder listing, pre-qualification and bid evaluation of construction companies

These duties are performed by the home office organization of the construction discipline that further continues serving the project during the construction period. This involves assigning construction personnel to site, assisting in planning the temporary field facilities, reviewing critical construction issues like heavy lifting, quality, and safety.

8 Engineering

Engineering can be defined as professional activities performed by applying scientific, technological, engineering, and mathematical knowledge as well as the accumulated engineering experience in design, development, and selection of materials, structures, machinery, devices, systems, processes, and other miscellaneous items for achieving the intended outcome economically, safely, and environment friendly.

ENGINEERING FOR PROCESS PLANT PROJECTS

A process plant can be regarded to be a complicated product of engineers from various disciplines including chemical, mechanical, civil, electronics, electrical, geotechnical, metallurgical and sometimes, other fields of engineering. Teams of engineers are involved in a process plant project from technology development or technology selection to completion of detailed engineering, procurement, construction, commissioning, and startup. Meanwhile, different professional fields other than engineering also take part in such projects, like finance, law, business administration, etc. However, the time spent by engineers for a project is almost always way beyond the hours by such others.

After buying the process package from a licensor or having it prepared by a competent entity, the overall engineering activities for a process plant project are fundamentally performed by the contribution of three different participants:

1. Engineering contractor(s) or sometimes, the engineering staff of owners
2. Vendors (manufacturers)
3. Construction contractor(s)

The first one above, the engineering contractor, is basically the guiding party of the entire engineering effort and spends the very major portion of the man-hours in engineering a process plant project. This party usually performs the basic or Front-End Engineering and Design (FEED) to be followed by detailed engineering, the latter often by another engineering contractor. As mentioned earlier, the objective of these activities is to establish the plant configuration, specifications, layouts, and all the other necessary engineering/design documentation, like job specifications, data sheets, requisitions, drawings, details, and material take-offs (MTOs) for correctly and timely supplying the material, carrying out fabrication, and construction and finally, the commissioning and startup. For licensed processes, the engineering performed by the licensor is usually limited to adapting know-how to the requirements of a particular project and sometimes include reviewing certain critical designs of the engineering contractor for compliance with the know-how.

Engineering by vendors comprises proprietary design for manufacturing equipment, packages, and systems, carried out with reference to requisitions or technical specifications prepared by the purchaser, which is usually the engineering contractor or sometimes the owner. Thus, the design and fabrication of such items are founded on the vendor's own know-how and carried out in accordance with the specific requirements of the project. During vendors' activities, the purchaser checks the inputs, outputs, compliance with requisition, and test results for ensuring that the requirements are fulfilled. The vendor drawings and data represent important set of information to be incorporated in the engineer's drawings for completing the detailed engineering by the engineering contractor.

The last type of engineering is the one carried out by construction contractors and usually comprises shop (manufacturing) drawings, certain installation details and schedules, like structural fabrication details, rebar schedules, MTOs, etc.

BRIEF HISTORY ABOUT ENGINEERING

The word, "engineer" is derived from the word "engine" which meant, several centuries ago, long before the industrial revolutions, a military machine while an engineer was the person constructing and operating such machines, such as a catapult. As a concept, engineering is believed to date even further back to ancient times to invention of wheel, pulleys, levers, etc. The pyramids of Egypt, Roman Aqueducts, Great Wall of China and all other ancient structures can be regarded as examples of ancient engineering. On the other hand, the word "engine" was derived from the Latin word "ingenium," meaning mental power yielding a clever invention.

As the design of civilian structures, such as bridges and buildings arose, the term civil engineering was started to be used for discriminating non-military engineering from the military one. The inventions by Savery by late 17th century and Watt in 18th century are considered to lead to the modern mechanical engineering. Electrical engineering is generally accepted to root back from the work by Volta, Faraday, and Ohm in the 19th century. In the same century, the studies made by Maxwell and Hertz are regarded as the starting point of electronics engineering. Development of chemical engineering is considered to take place during the industrial revolutions.

Today, engineering is a broad field of professional work, categorized into a number of disciplines, such as chemical, civil, mechanical, and electrical, for various different major areas. Until lately, engineering education is used to be on single discipline in one of the above fields; however, there are lately some combinations like mechatronics, as well as double major or major-minor opportunities for engineering students who can accordingly make registration to a second engineering education program.

In modern-day professional life, an engineer often attains sort of multidiscipline capability due to the unavoidable need for continuous learning and accumulation of experience on all aspects of a particular field of work along the path of career in a company.

TYPES OF ENGINEERING PACKAGES

Engineering performed by engineering contractors falls mostly in one of the below categories, each of which serves for a specific stage during implementing process plant projects. These categories are listed in the order from early to late stages of the entire engineering:

- Process engineering package
- Basic engineering
- FEED
- Detailed engineering

In other words, the start of engineering with process package should progressively move to the end and finish with detailed engineering by which material supply, manufacturing, construction, and plant operation of the project are facilitated.

As mentioned earlier in this book, for projects of limited size and complexity, some of these services are sometimes rendered by the project owners themselves, depending on availability of the in-house know-how and capability of the staff.

PROCESS ENGINEERING PACKAGE

A process engineering package is prepared at the very beginning of a process plant project, either by a licensor (for licensed processes) or by the owner or sometimes by an engineering company (for open art processes). The package contains basically information on the process technology, including the following:

- Basis of process design and calculations
- Material and energy balances, process flow diagrams (PFDs)
- Preliminary process piping and instrumentation diagrams (P&IDs)
- Process equipment data sheets
- Process instrument data sheets
- Fluid list
- List of utility requirements
- Process description
- Control philosophy
- Basic materials of construction (MOC)
- Design basis for safety/relief devices
- Identification of health and environment hazards
- Specific process requirements relevant to layouts and elevations

The process package forms the basic guide for the subsequent engineering stage and needs to be expanded in the next step to either basic engineering or directly to FEED before starting with the detailed engineering.

BASIC ENGINEERING

Basic engineering usually covers the basis of design for the entire engineering of the plant including the following:

- Basic engineering considerations for all engineering disciplines, including instrumentation and controls, electrical, piping, rotating and static equipment, other mechanical systems, civil, structural as well as process and process safety

- Extended process engineering package accompanied by process issue of P&IDs and basic information on utilities

For process units, patented by licensors, a thorough review of the license information is normally required as part of basic engineering in order to identify any omissions or mistakes in the license package, so that such deficiencies are timely corrected in contact with the licensor. For engineering disciplines other than process, the package usually contains the following:

- Plot plan and equipment layouts (preliminary issue, usually based on a piping scope layout)
- Piping material specification
- Preliminary equipment list
- Design basis for mechanical equipment (major static and rotating equipment items)
- Design basis for instrumentation and automation with a preliminary instrument list
- Civil/structural design basis with preliminary sizing of major structures, preliminary design for buildings (architectural, civil, mechanical and electrical works)
- Design basis for electrical works including preliminary electrical single line diagrams and preliminary electrical consumers list
- Design basis for fire protection and other safety-related systems
- Design basis for sewer systems and waste treatment

The package often refers to a geotechnical report and site mapping prepared earlier by others for the plant site.

FRONT-END ENGINEERING DESIGN

FEED is not necessarily a next step, but is an alternate to basic engineering package with the purpose of providing more detailed information than the basic engineering. In other words, the difference between the two is mainly a higher level of elaboration as well as more project documentation presented in the FEED compared to basic engineering, in order to make smoother and faster transition to detailed engineering. It is the front end of the remaining efforts not only in engineering, but also in other functions, like equipment and material supply.

From an owner's perspective, FEED package is the most typical package included in the Request for Quotation (RFQ) for Engineering-Procurement-Construction (EPC) bids, as it allows bidders to develop an EPC grade cost estimate, with a relatively much less effort and risk. Additionally, issuing RFQs based on a FEED package would greatly increase the chances of receiving good and comparable proposals. For additional details on cost estimates, please refer to Appendix 1 – Cost Estimating. Accordingly, vendor lists as well as requisitions for long delivery items are usually prepared during the FEED. It represents a more descriptive level of engineering as all job specifications for engineering disciplines are made ready and so, the project-specific standards are set forth together with selecting the codes and standards to be applied during final engineering stage, the detailed engineering. Additionally, FEED packages may contain well-established execution plans including project schedules for the rest of the project. It is also possible to make many important assessments and reviews, like quantitative risk assessment (QRA), Hazard and Operability (HAZOP) review, layout review, constructability review, etc., at the end of the FEED. In most cases, an updated cost estimating is prepared and submitted to the owner for re-evaluating the decision to proceed with the next steps of the project. Based on the updated cost estimate, it is also possible that a project may get suspended or cancelled at that stage.

Detailed Engineering

Upon award of an EPC or EPCm contract based on the owner-approved FEED, the detailed engineering activities are started by the selected contractor. However, for small or simple projects that are often under Engineering Services-type engineering contracts, a separate prior engineering phase, as basic engineering or FEED is not necessarily performed, and instead, engineering can be carried out seamlessly to cover all the engineering phases with reference to a process package.

Approval of the prior engineering package before starting the detailing, serves to freezing the principal design parameters and basic plant configuration that have been worked out until then. This, in turn, helps minimizing costly and time taking step backs and re-doings during detailed engineering. If changes in the frozen basic design or FEED are still requested by the owner or required for any other reason, then cost and schedule impacts of such changes should be estimated in comparison to the schedule and the budget valid at the time of freezing the design. Such changes will require to be verified by the owner if to be compensated, either with or without schedule extension. For more details about change order management, please refer to Chapter 18 – Contract Administration.

A detailed engineering constitutes the largest production phase of the entire engineering efforts in terms of man-hours and the number of engineering deliverables. So, after having the basic design or FEED frozen, way more deliverables are produced by loading much more manpower to the project.

Detailed engineering is performed basically to serve the physical realization of the project at the jobsite for having the plant built and started up subsequent to carrying out the purchasing, construction, commissioning, and startup. Almost all requisitions for supply of equipment and materials are prepared (or at least finalized) during the early periods of this stage, with some exclusions for long delivery items that need to be delivered earlier for meeting the plant completion date. Detail engineering specifies the material supply, manufacturing, and construction information to the highest level of detail sufficient for enabling vendors and construction contractors to proceed with their works without any ambiguities on what to do.

Detailed engineering for a process plant leads to production of hundreds to thousands of deliverables (i.e., drawings, specifications, etc.), depending on the size and complexity of the project. If the project is for a complex multi-unit process plant, like a refinery or a petrochemical complex, then the number of deliverables becomes extensively high in relation to the number of units as well as complexity of each unit. It is not unusual to have several thousands of piping isometric drawings only for one of such sizable process units. If the number of documents produced by all disciplines is added up, a very high number, like a couple of ten thousand will not be surprising. For such sizable projects, it would not be extraordinary to spend hundreds of thousands of man-hours only for engineering.

Detailed engineering is made up of complicated bundles of activities, needed for specifying the detailed plant set up and for facilitating material supply, construction, and operation of industrial plants. It consists of calculations, reports, designs/drawings, data sheets, requisitions, specifications, and take offs for realizing the entire project work. Without properly rendered engineering, it is unlikely possible to avoid facing difficult to solve problems, as mentioned in Chapter 7 – Project Development during Home Office Services.

Detail engineering can also be defined as producing very specific design details in line with applicable codes, standards, and regulations by incorporating also the data of the actually supplied material and equipment for achieving "for construction" level of documentation to be applied at the construction site.

ENGINEERING DISCIPLINES

Engineering and design for process plants require collaboration and coordination of a variety of engineering disciplines and integration of different trades of design in the plant design work. Process plant projects, that are usually more complicated than other industrial plant projects, necessitate notable amount of process engineering work, major part of which is performed during the beginning of the project. Process engineering then continues its role by making follow-up, attending engineering reviews, writing operating manuals, and participating to start up and performance tests.

Coordination among different engineering disciplines is provided normally by project engineering coordinators, project engineers (PEs) and design-drafting coordinators. For process plant projects dealing with liquids and gases, the highest portion of the detailed engineering man-hours is usually consumed by piping discipline. For solid-fluid treating plants, the top man-hour consumer mostly becomes the mechanical equipment groups that design the solid and gas transportation and separation system components including ducts, chutes, rotary feeders, feeding vibrators, conveyors, elevators, centrifugal separators, baghouse filters, etc. When working on power plant and metallurgical plant projects, electrical engineers usually require more man-hours compared to what they would normally need for other type of plants. Obviously, the above are general trends that highly depend on the type of process as well as the scope and size of the plant.

Another important point, worth mentioning here, is the role of design and drafting that has to go hand in hand with engineering during the engineering phases of process plant projects. Engineering mostly serves by making calculations, specifications, selections, verifications, follow-ups, and problem solving while design takes its role by devising layouts, set ups, configurations, and fabrication and construction details to be incorporated in the drawings. In professional engineering companies, each specialty discipline usually has its own designers and drafters working with the engineers of that department. The foregoing arrangement does not normally apply to process and project engineering that have much less drawings to work out. These disciplines have their drafting work handled by another discipline, like piping.

Duties of the engineering disciplines taking part in process plant projects are outlined in the following sections.

PROCESS ENGINEERING

Process engineers are the professionals who deal with engineering, design, checking, and controlling of industrial processes. They are mostly chemical engineers as mentioned in Chapter 1 – Preamble.

Process engineers play a crucial role in process plant projects by defining the basic plant configuration and by specifying in detail the process requirements related to equipment, instrumentation, utilities, control philosophy, plant operation and performance as well as process safety and environmental care. The engineering and design deliverables by process engineers are then distributed to other disciplines, primarily to the project engineering, piping, mechanical and instrumentation-automation engineering. The dispatch of the process information to other departments acts as switching on detailed engineering and design activities by these disciplines. For non-process-type industrial plants, the role of process engineers is limited to utilities and utility distribution.

As mentioned before, the basic process technology is supplied in one of the following ways:

- Know-how (license) from licensors for patented units
- Know-how possessed by the owners through their experience on running and developing similar plants (mostly after expiry of the original license they once possessed)
- Process knowledge and experience accumulated by engineering contractors (for unlicensed/open art processes)

Even when the process know-how is received from a licensor, the process engineers still have to do critical process engineering activities, which basically consist of making thorough process checks for highlighting and clarifying any possible omissions and mistakes in the license package delivered by the licensor.

Main duties of process engineers can be summarized as follows:

- Review and set up the entire process configuration of the facility under design, including process units, utilities, storage, control systems, etc.
- Prepare flow sheets.
- Perform material and energy balances; calculate flows, pressure drops, heat transfers, plant performance, equipment and piping sizes; make process simulations.
- Specify process and equipment items, utility systems, instrumentation, and controls.
- Follow-up detailed engineering activities in relation to the process designed.

- Make technical evaluations and follow up for major process equipment along purchasing activities.
- Prepare operating manuals.
- Participate in startups and test runs.

Main categories of project documentation prepared by process engineers can be summarized as follows:

- Basis of design
- Process and utility calculations, simulations
- Flow sheets
- Process equipment and process instrument specifications (data sheets)
- Utility design and summary
- Control philosophy
- Operating manuals

Basis of design normally covers process description, battery limit conditions, main physical and chemical properties of materials including feedstocks, products, by-products and catalysts, process conditions and constraints, applicable codes, standards and regulations, utility requirements and other information that has effect on the process design.

Flow sheets prepared by process engineering are the following:

- Process Block Diagrams (Block Flow Diagrams: BFD)
- PFDs
- P&IDs
  - Process P&IDs
  - Utility piping and instrumentation diagrams (UIDs, sometimes called utility P&IDs to supplement the ones prepared by utility vendors)
  - Distribution P&IDs (distribution of the process and utility fluids and interconnecting)
  - Symbols P&ID (or P&ID Legend)

PFDs and P&IDs are of major tools in understanding and following up the process and operation of the plant.

BLOCK FLOW DIAGRAM

Block Flow Diagrams (BFDs) are drawn for a single process and consist of a series of rectangular blocks representing different equipment or unit operations. Each block represents a function/operation and may consist of one or several items of equipment connected by input and output streams. Major flows are shown by arrows connecting the blocks. It helps in conceptualizing processes.

Plant Block Flow Diagram is a type of BFD used for illustrating complete chemical complexes composed of multiple units with each block representing a complete process, so shows a complete picture of the entire complex and interaction of the processes.

Process Flow Diagram

A higher level of process information is presented in PFDs compared to BFDs. Formats and contents of PFDs may change from company to company.

A PFD, for most users, indicates the following:

- Process topology (position of and interaction between equipment items and flow streams)
- All major pieces of equipment (but not minor items) with specific symbols
- All main process flow streams (but not all piping lines)
- Process conditions in form of stream information or flow summary, such as flow rate, composition, temperature, pressure, and vapor fraction of streams
- Utility streams supplied to major equipment for a major process function
- Basic control loops (only), showing the control during normal operation mode

PDF further includes the following:

- An equipment number for each equipment item shown
- A stream number (not a line number) for each process stream shown
- Equipment information

PFDs are prepared and used by process engineers. Apart from drafting by hand or by 2D computer-aided design (CAD), this is the type of flow diagram sketched by using the process simulation software before starting process simulations. These diagrams are also used by others basically for visualizing and understanding the process.

Piping and Instrumentation Diagrams

Piping and Instrumentation Diagrams P&IDs are prepared by process engineers, as the first (process) issue, which is then taken over usually by PEs for making the subsequent issues. Furthermore, P&IDs are:

- Commented and used by instrument and automation (control) engineers.
- Used in preparing Distributed Control System (DCS)/ Programmable Logic Controller (PLC) screen mimic displays.
- Represent a major tool for mainly the piping discipline.
- Used as the most important reference during health, safety, and environment (HSE) reviews.
- Used by commissioning engineers during commissioning and startup.
- Used by plant engineers for getting trained on operation and maintenance as well as trouble shooting.

P&IDs are one of the most key design, operation and maintenance documents of a process plant. They show the following:

- All lines of piping, ducts, and chutes with a line number indicating also diameter, schedule, material of construction and insulation

- All equipment with tag numbers and equipment information table
- All instrumentation and controls with symbolic representation of connections
- All utility connections identified by numbered boxes

Indications for certain important specifications, like piping material and insulation are shown on P&IDs, mainly within the line number. Instrumentation is indicated by circular flags or balloons that denote instrument tags.

Since P&IDs show extensive amount of engineering information, each drawing is usually limited to show only a few of the equipment items on each drawing. Lines may enter from and leave to other P&IDs.

Other synonyms of P&ID are Engineering Flow Diagram (EFD) and Mechanical Flow Diagram (MFD).

A sample P&ID, sketched for condensation of a hydrocarbon vapor and pumping the condensate is shown in Figure 8.1.

PROCESS SPECIFICATIONS (DATA SHEETS)

These documents show process engineering information about equipment and instrumentation, mostly in tabular form. They specify the requirements from process engineering perspective and are then used by other engineering disciplines to form their design documentation by putting in additional requirements specific to each such discipline.

The duties performed by process engineering in process plant projects are shown in Figure 8.2 as a summary of the main activities.

Please note that the above figure is a simplified work flow diagram, prepared by omitting interdisciplinary information exchange for the sake of keeping the figure simple. Also omitted are the technical bid evaluations and vendor information reviews, which are separately shown later in this chapter in Figure 8.14 as applicable to all the engineering disciplines.

PROJECT ENGINEERING

A PE is a member of the project team, responsible for the proper execution and integration of the entire engineering of the project. In that respect, he or she is at the centroid of the engineering activities as a coordinator and a help desk for the other engineering disciplines involved. Such responsibilities and representations are based on PE's close follow-up of the development of the engineering and are also extended to construction phase once the PE is assigned to site with the title of field project engineer. In many engineering companies, project engineering has dual functions in a project. These are as follows:

1. Coordination of the project
2. Handling the key project documentation

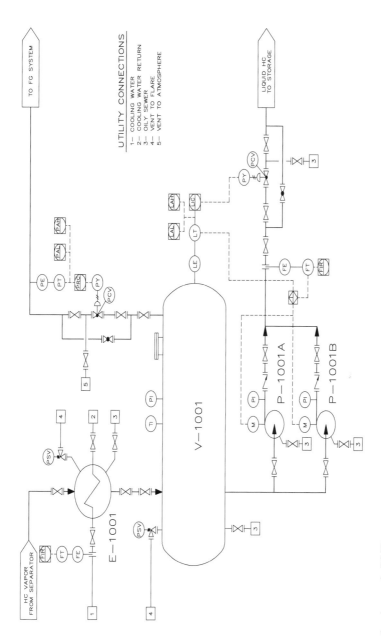

FIGURE 8.1 A sample P&ID.

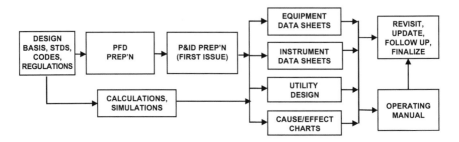

FIGURE 8.2　Simplified work flow diagram for major process engineering activities.

COORDINATION OF THE PROJECT

A prime PE function is the project coordination, which can be outlined as follows:

- Ensuring timely flow of project information among different parties involved in the project
- Handling project correspondence both, inside the project team and with external parties on behalf of the project manager
- Setting up the project filing system and ensuring proper filing is maintained
- Participating in project meetings
- Writing and distributing minutes of meetings
- Reviewing project documents issued by others and distributing, when required, to the team
- Acting as a reference point or help desk for other disciplines to give support by advising about project requirements and developments when requested
- Preparing project coordination procedure and basic engineering design data
- Assisting PM on project matters, including participating in the meetings and correspondence with the owner, vendors, etc.

Figure 8.3 indicates the major coordination functions of project engineering. The boxes in lighter color show external interfaces of an EPC/EPCm contractor.

When a project engineer is assigned to a construction site for acting as the field project engineer, he or she reports to the manager of the construction team and performs usually the following duties:

- Provide link with home office on engineering and drafting matters.
- Perform link with owner's engineering representatives at site.
- Assist site supervisors in commenting on method statements by contractors.
- Resolve at field any discrepancies that may be identified in drawings and specifications.
- Supervise the work of field drafters, if any, on site.
- Assist in interpreting scope coverage and requirements of construction contracts.
- Update P&IDs to incorporate latest changes in the field.
- Assist site planner in interpreting the engineering about the remaining work.

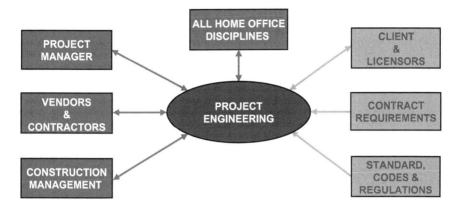

FIGURE 8.3 Main coordination functions performed by PE.

- Perform flow check, as required, with reference to P&IDs.
- Define hydraulic test circuits.
- Ensure "as-build" mark ups are properly made all along the construction work.
- Assist the construction management team in compiling the field records and certificates.

HANDLING THE KEY PROJECT DOCUMENTATION

Duties of project engineering on key project documentation cover preparation and updating of the following:

- P&IDs (mechanizing and updating after taking over from process engineering discipline)
- Line list
- Equipment list
- Fluid list
- Tie-in list
- Area classification
- Requisitions for packages and miscellaneous specialty items
- Construction requisitions
- Job specifications
- Technical bid evaluations and vendor information reviews for PE-prepared requisitions

P&IDs are prepared by the process engineers, as mentioned in earlier chapters. Once the first issue (the process engineering issue) is made, then the responsibility for detailing and updating these diagrams passes to project engineering in some companies, though may still be kept by process engineering discipline in some others. The main tasks carried out by project engineers on P&IDs are first, the mechanization and then, updating in line with the developments along the life of the project. The mechanization covers a number of activities including the detailing of the P&IDs for enabling use by other engineering disciplines. This involves adding in line numbers (that cover indication of piping material specification and insulation as well as the line size), arranging and proper labeling of the "in" and "out" streams, tabulating on these drawings the size/capacity, operating and design conditions of equipment. The updating of P&IDs includes incorporating all later process design revisions (PDRs) by process engineers and comments by instrument discipline, piping discipline and the owner, as well as incorporation of actual vendor information. Additionally, some types of P&IDs are mostly prepared by PEs. These are dismantling P&IDs (showing the equipment, lines and/or instrumentation to be demolished in an existing unit), distribution and interconnection P&IDs.

While working on the P&IDs, PEs prepare and take the updating responsibility also for the line list, equipment list, and the fluid list. This later follows incorporating actual vendors' information as soon as available. Another document prepared by project engineers is the tie-in lists that are needed for projects inside an existing plant where certain piping connections are to be made to the tie-in points of existing lines or equipment.

Producing and updating area classification is usually one of the responsibilities of project engineers, though such responsibility might have been given to other engineering departments, like mechanical or electrical in some engineering companies. Based on the code applied in the project for classification of hazardous and non-hazardous areas, the project engineer marks up on layouts and sections the extent of each area classes and have that information drafted as area classification drawings. These drawings are mainly referred while selecting the classification of electrical items of the project including electrical/electronic instrumentation, as well as when carrying out any plant modification.

In a process plant project, system packages designed, manufactured, and often erected by vendors may be needed as part of the new facilities. Some examples of these packages are demineralized water, steam generation, inert gas, refrigeration, waste water treatment, etc. These packages are normally complete units, containing various mechanical, piping, instrumentation, structural and electrical components. Accordingly, for preparing such requisitions that involve multidiscipline engineering, it is often the PEs who undertake the duty with technical support from relevant specialty engineering disciplines.

Another document having a multidiscipline nature is the construction requisition that is again prepared mostly by PEs with similar support from the other disciplines. Even when the construction requisition seems to cover a single trade of work, like site grading or civil works, it often includes some other types of work, such as installing grounding network, or underground piping, which would better to be done during excavations and landfill.

Job specifications are originally the engineering standards, often included in quality systems of engineering contractors or owners. These are generated from those standards by adjusting such standard documentation to fit to the specific project requirements. Preparation of job specifications is often a task performed by PEs for most of such specifications, obviously with support from specialty engineering disciplines as required. These specifications are mainly used for

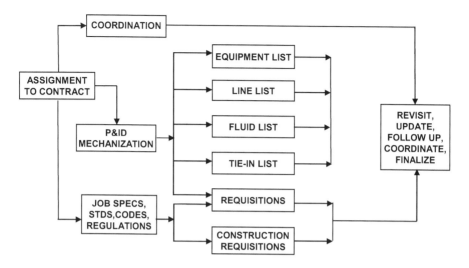

FIGURE 8.4 Simplified work flow diagram for major project engineering activities.

supplementing the inquiries and orders by attaching to inquiry packages as well as to subsequent POs/contracts.

A summary of the home office activities of project engineering discipline is given in the work flow diagram in Figure 8.4.

The above figure omits interdisciplinary information exchange for the sake of keeping the figure simple. Technical bid evaluations and vendor information reviews, which are also omitted, are separately shown in Figure 8.14.

INSTRUMENTATION AND AUTOMATION ENGINEERING

The discipline is also named as "instrument engineering" or "instrument and control engineering." It will be referred to as "instrument engineering" in the following paragraphs.

Instrument engineering expertise has mainly two subdivisions, namely:

1. Field instrumentation
2. Automation/Controls

The field instrumentation covers a wide variety of instrument items, like measuring points (e.g., thermowells, orifice flanges), indicators (e.g., temperature, pressure, level, flow, specific gravity, weight, viscosity, position), transmitters, local controllers, local alarms, local recorders, a variety of valves (like safety relief valves, control valves, solenoid valves, etc.) and other miscellaneous instruments including analyzers.

Instruments are connected to piping, ducts, valves, and equipment through certain sets of mechanical (and often continued with electrical) connection components. Details showing these connections are termed as hook-ups. Instrument engineers require reviewing and commenting to the project drawings, such as P&IDs, which show instrument connection points. In this respect, P&ID instrument comments and instrument vessel comments are one of the standard tasks performed by this discipline. There may also be instruments on panels as well as at back of panels that are either in the control room or inside a unit as local control panels.

Automation systems are computer-based monitoring and control systems that consist of a DCS and/or PLC and SIS (Safety Instrumented Systems, formerly referred to as ESD, Emergency Shut Down systems) connected to the field instrumentation through cabling, junction boxes, and panels. Buildings may as well be equipped with BMS (Building Management System), alternately termed as BAS (Building Automation System) that can be used to control parameters and operating modes inside the buildings, for mechanical and electrical systems such as HVAC (heating, ventilation and air conditioning), power, lighting, firefighting, access control, CCTV, telecommunication, and security systems.

Powering of instruments are either electrical (DC or AC) or pneumatic. Electric power for instrumentation and automation is usually fed through UPS (uninterrupted power supply) that keeps supplying power usually for some fraction of an hour on a black out. As certain portions of the facilities usually constitute critical loads that require to be operated during any extended power outages for plant safety, such loads are also connected also to DGs (diesel generators). UPS and DG constitute emergency power generation system of a facility which is crucial for safe shut down of the process unit during an extended power outage. Pneumatic systems are powered by instrument air systems equipped with compressors, dust and oil filters, air driers, air receivers, pressure regulators and instrument air piping, tubing and fittings.

Instrument engineering practically starts with the internal issue of P&IDs within the engineering organization by the process discipline. Before making the first official issue, P&IDs are submitted to the instrument engineers who review these drawings to assign instrument tags as well as give comments in line with the control philosophy and instrument job specifications. Meanwhile, preparation of instrument list and instrument requisitions is started by referring to the process instrument data sheets. During this stage, estimating the number of input and output signals (both digital and analogue) can be done to generate the I/O list. The number of I/Os (inputs and outputs) is one of the major parameters to be included in the requisition of an automation system.

Field instrument requisitions indicate process data, range, type, material, and other particular requirements relevant to the requisitioned instruments in a data sheet format. However, requisitions for DCS, SIS, and BMS are rather voluminous due to more complicated nature of these systems. Requisitions are transmitted to the purchasing department as soon as they are made ready for inquiry. Instrument bulk material requisitions for instrument electrical and instrument piping bulk items can be prepared much later during the project. Such tasks need certain level of development in instrument design for preparing MTOs with reference to instrument layouts and instrument installation details.

Instrument piping specification is prepared with particular reference to the piping material specification of the project for achieving material compliance. This type of specifications is actually produced by materials engineers and used throughout the project in selecting instrument and piping bulk material. Other instrument specifications are prepared mostly in collaboration with the project engineer, who has extensive knowledge of the project requirements while instrument engineer contributes to the work with regard to technical content specific to instrumentation.

Instrument layouts can be prepared at relatively later stages of the instrument design after equipment and piping layouts are developed by other disciplines to a good level. Similarly, instrument cable layouts require development of these layouts as well as the interconnecting structures like pipe racks, for properly coordinating with electrical, piping and structural disciplines the routing of cable trays and conduits. In general, routing of instrument cabling is designed to run close to electrical cabling in order not to unnecessarily occupy other spaces inside the plant, but still with certain distance in-between or with proper separation for eliminating magnetic field generated by power cables, which could affect instrument signals. Hence, it is a general design trend to run instrument cable trays along the pipe rack, but on a different level from electrical cable trays/ladders or sometimes from the other side of the rack.

Loop wiring diagrams are schematic indications of connections and loops formed for field instruments located at the front and back of instrument panels. Most of other drawings for control and automation, like ladder diagrams, logic diagrams, mimic drawings (computer screen page configurations), etc. are usually prepared by vendors with reference to the P&IDs and instrument drawings attached to the requisition. Vendor-proposed designs are obviously reviewed and commented by instrument engineers.

Main categories of documentation prepared by instrument engineering discipline for process plant projects are the following:

- Job specifications (support to PE when prepared by PE)
- Instrument list (also named as instrument schedule)
- I/O list
- Instrument sizing calculations
- Requisitions for field instruments
- Requisitions for DCS, SIS (ESD), BMS
- Instrument piping specification (with support from materials engineer)
- Instrument installation details
- Loop-wiring diagrams
- Junction box drawings
- Instrument layouts
- Instrument cable layouts
- Control room layout
- Instrument bulk MTO
- Instrument bulk material requisition
- Technical bid evaluation and vendor information review

A summary of the main instrument engineering activities is given in Figure 8.5 in a work flow diagram format.

Interdisciplinary information exchange is omitted in Figure 8.5 for keeping it simple. Technical bid evaluations and vendor information reviews, which are also omitted, are separately shown in Figure 8.14.

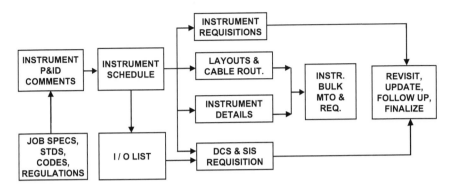

FIGURE 8.5   Simplified work flow diagram for major instrument engineering activities.

PIPING ENGINEERING AND DESIGN

Piping engineering discipline mainly performs piping design, pipe stress analysis, pipe supporting, and piping MTO. For process plants that treat fluids, such as refineries, petrochemical plants and many chemical units, piping discipline has a very major role in the entire detailed engineering phase of a project. This is due to the following:

- The portion of the piping man-hours in engineering of such projects is normally much more than the others
- Piping activities dominantly drive the entire detailed design of the project

The above aspects are obviously related to having such plants congested with piping with many lines entering and leaving equipment, making by-passes and loops, running within the plant in bundles of piping on pipe racks, extending to other units by running along pipe racks and pipe ways. Hence, many lines coming in and going out in various directions and different elevations result in having the piping discipline spend quite a lot of time in laying out equipment and piping simultaneously for a good design satisfying all process, operational, maintenance, and safety requirements. Some of such requirements that piping designers should always be careful about are as follows:

- Minimum connection distance, sloped pipe, no pockets, etc., when so specified on the P&ID
- Correct configuration for in-line instruments and sampling points
- Sufficient space for thermal loops as well as for insulation of hot or cold pipes and equipment
- Compliance with vendor design (e.g., allowable loads on equipment flanges)
- Access to valves, instruments and certain equipment parts for operation, testing, and maintenance
- Space for dismantling equipment parts (e.g., heat exchanger tubes)
- Ease of access to the unit during operation, maintenance, and emergency
- Proper supporting in order to eliminate major changes during stress analysis
- Provisions for other supporting structures and cable raceways

- Properly designed and located vents and drains for startup, operation, and shutdown
- Provisions for future additions of new lines
- Other process/operational requirements like, pump suction lines, compressor re-cycle lines, upstream and downstream lines of flow meters and desuperheaters, etc.
- Flow checks and clash checks

The heavy role of piping discipline in fluid processing plant design drops significantly for solid processing plants. The latter type has the material movement accomplished through mechanical items such as conveyors, elevators, chutes, rotary feeders, dust collection system with exhaust fans, hoods, cyclone separators, ducts, and filter bags. Consequently, the size and complexity of piping design is relatively much lower and so the driving role in design passes to mechanical engineering groups.

In process plant engineering companies, piping discipline is not normally involved in plumbing design which constitutes small size sanitary and HVAC piping mostly for buildings. This type of work is usually handled by a mechanical engineering discipline responsible for building mechanical design. Similarly, engineering and design of drainage and sewer systems are usually out of piping design scope and are covered mostly by civil engineering discipline.

Piping design requires to be based on P&IDs and is started as soon as the first issue of these diagrams is made. The first design study is usually the so-called "scope layout," objective of which is to make preliminary layout sketches to be the basis for the piping layouts and plot plan as well as the designs by all other disciplines. Such a study targets achieving the best possible design solution that will eliminate major modifications during the rest of the project and so, requires a very considerate and clear foreseeing of the eventual plant as best as possible by taking into account all process, operational, maintenance, and safety requirements. Scope layout basically shows the following:

- Equipment with the best estimated shape and dimensions (estimating need due to unavailability of vendor drawings at such an early stage)
- All major piping lines and connections including space for future additions
- Preliminary configuration of major structures that occupy considerable space, especially pipe racks and platforms
- Allocations for cable raceways, future expansions, etc.

As soon as a good scope layout is obtained, piping discipline can start preparing the plot plan, piping layouts and later, the piping isometric drawings with reference to P&IDs, line list and piping material specification. Meanwhile, with the piping scope layout made available, civil-structural discipline can commence its design for major structures, such as pipe racks.

A Plot plan is a general layout of the entire plant showing major outdoor equipment, structures, roads, buildings and other major items, such as plant entrance/exit gates and fencing. It is one of the most important key drawings of a process facility for all the engineering and construction disciplines. A sample plot plan showing general layout of an oil storage and distribution terminal is illustrated in Figure 8.6.

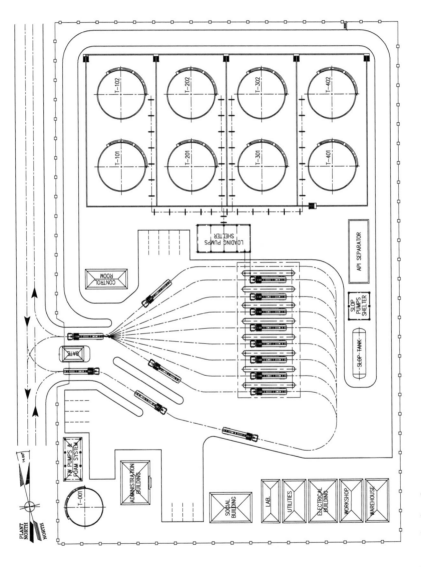

FIGURE 8.6   A sample plot plan.

Layouts are plan views of certain areas and/or systems of a plant showing dimensions and prepared on certain scales as well as for different elevations as necessary. This is the same for piping layouts and shows the piping of an area with all in-line components, structures, connections, dimensions, and elevations. A sample piping layout for a pump area is shown in Figure 8.7.

Piping isometric drawings are piping details used for piping MTOs, fabrication, and erection. Preparation of piping isometrics (also called "isos") passes through the following stages:

- Isometric sketching (isometric modeling in 3D design)
- Isometric checking
- Flow checking
- Stress-support checking
- Ready as "IFC" (Issue for Construction)

In some parts of a project, "piping general arrangements" may be prepared instead of piping layouts plus isometrics. The former drawings are further detailed piping layouts, containing sectional views, and other necessary details to compromise unavailability of isometrics. The addition of details serves providing piping fabricators/erectors sufficient information to do their work. Such an approach is usually selected for lowering piping design man-hours; however, for medium- to large-size projects as well as tight-scheduled jobs, this is not the preferred way, except for almost straight run portions of the piping, like the pipe rack or pipe way piping.

A sample piping isometric drawing is indicated in Figure 8.8.

For projects designed using 3D software, CAD operators model the equipment first, and then the piping, structures and the cable raceways. After modeling is completed, it is possible to extract piping layouts, isometrics, and MTOs from the system that has been pre-loaded with the necessary project standards including piping material specification and the standard piping details. One of the advantages of using 3D systems is the feature of clash checking. This helps eliminating clashing mistakes of pipes with other pipes, equipment, structures, and/or cable raceways. Another important advantage brought by 3D design is the option of making "walk through" type of model reviews as well as the possibility of taking views from any different levels and directions. Such model reviews sessions are usually organized with the participation of the owner at certain design stages, like 60% progress and 90% progress and help collecting comments from the participants. Figure 8.9 shows a 3D model of a refinery process unit.

A main engineering task of the piping discipline is the stress analysis that verifies the integrity and flexibility of the piping and support design with reference to static and dynamic loads exerted on the piping. Such loads may result from the following:

- Thermal expansion (or contraction)
- Internal and external pressures
- Fluid flow (friction, drag force, water hammer)
- Gravity (weight of the pipe filled with process fluid and hydro test fluid)
- Occasional natural forces (seismic and wind)

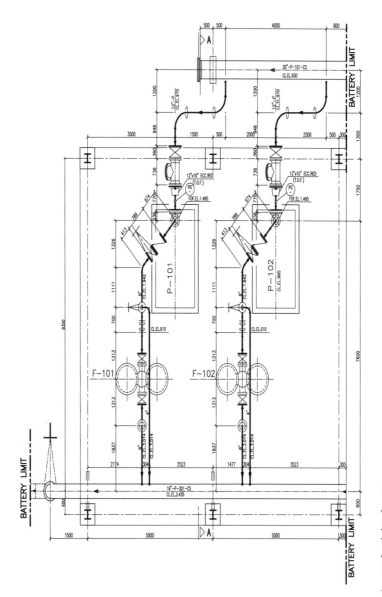

FIGURE 8.7   A sample piping layout.

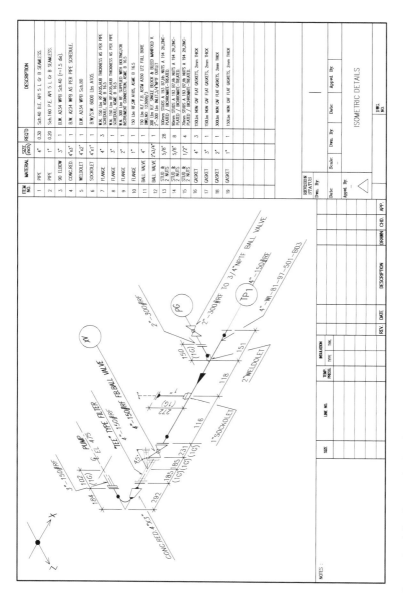

FIGURE 8.8 A sample piping isometric drawing.

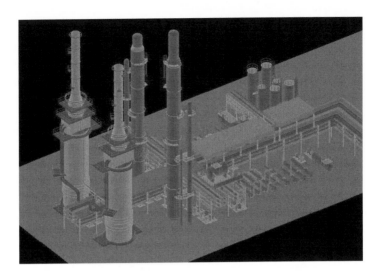

FIGURE 8.9   3D model of a process unit.

Pipe stress engineers review the design with respect to pipe configuration, nozzle loads, supports and hangers, and check if allowable pipe stress limits are exceeded under possible situations (e.g., hydro testing, operation, heavy wind, earthquake). These reviews require extensive calculations and are done by using professional stress analysis software.

Piping MTO is another work carried out by the piping discipline. It is also a critical and extensive work due to having usually so many different types of piping items in various different MOC and a variety of specifications in terms of size, pressure rating, and connections. All these are to be counted at each updating of piping drawings. Basic parameters used for classifying piping material items in an MTO are the following:

- Type of item (pipe, fitting, flange, valve, gasket, stud bolts, etc.)
- Material of construction (various types of carbon steel, alloy steels, stainless steels, titanium, plastic, etc., to be identified with reference to the piping material specification)
- Pressure rating (in pounds as per North American system like 150#, 300# and nominal pressure in European system like PN20 for valves, fittings and flanges)
- Thickness (in various schedules such as Sch.40 and Sch.80, in North American system and in mm based on outside diameters in European system)
- Diameter (for most items, as nominal pipe size, "NPS," in North American system like 12″ and nominal diameter, DN, in mm in European system)
- Length for pipe, mostly as 6 or 12 m
- Connection or ends (flanged, butt weld, socket weld, threaded, beveled end, plain end, etc.)

Some of the main parameters above need to be further divided into sub-types in the piping MTOs. Hence, the "type" is normally to indicate also the following:

- Pipes (uncoated, internally coated or externally coated with different coating materials and coating thicknesses)
- Fittings (elbows, tees, couplings, half couplings, unions, weldolets, etc.)
- Flanges (RF: Raised face, FF: Flat face, and RJ: Ring joint)
- Valves (gate, globe, check, ball, etc., also with type of connections as flanged, welded, wafer, etc.)
- Gaskets (various types of gaskets also based on the type of flanges the gasket is to fit in)
- Other in-line piping species (like strainers, blinds, sight glasses, etc.)

Material of construction is to be properly specified by referring to a specific material standard because indicating only carbon steel or alloy steel, etc., does not sufficiently describe the material of construction as there are a variety of different carbon steels, alloy steels, plastics, etc.

All these parameters are fully covered in a piping material specification, arranged in form of classes for each different piping service, classified according to the type of fluid and operating conditions in the plant. Hence, this brings in thousands of different piping items to be dealt with all along the project. It has lately become a common practice to perform MTO continuously during the entire design stage; however, updates of these lists are issued to procurement at certain stages, usually named as 1st MTO (30% of material, mostly valves taken off from P&IDs), 2nd MTO (60% or more depending on total number of MTO issues decided to be made), 3rd MTO, etc. Making such partial or incomplete issues of the MTO during the progress of the project is quite essential for making available at site for the construction stage always sufficient amounts of piping material.

A brief outline of major piping activities can be made as follows:

- Piping job specifications (support to PE when prepared by PE)
- Piping scope layouts and piping layouts
- Plot plan
- Piping general arrangement drawings
- Piping isometric drawings
- Model reviews
- Piping stress analysis
- Pipe supports
- Piping MTO
- Piping bulk material requisitions
- Technical bid evaluation and vendor information review for piping items

A summary of the main activities of this discipline is given in Figure 8.10, again in a simplified work flow diagram form, omitting the interdisciplinary information exchange for keeping the figure simple.

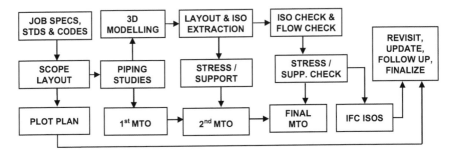

FIGURE 8.10 Simplified work flow diagram for major design activities by piping discipline.

EQUIPMENT/MECHANICAL ENGINEERING

Concerning mechanical works of process plant projects, mechanical engineering disciplines involved basically work on static equipment, rotating equipment, material handling, and building mechanical works. From an organizational perspective, performance of the aforementioned services is usually carried out by a couple of separate disciplines, often named in a variety of ways in different engineering companies. Material engineering duties are sometimes included in one of those disciplines, though it is more of a metallurgical expertise than of mechanical. The list of main expertise covered by these disciplines is as follows:

- Static equipment
  - Vessels
  - Heat exchangers
  - Heaters
  - Equipment platforms
- Rotating equipment
  - Pumps
  - Compressors, blowers, fans
  - Turbines
- Material selection
- Building mechanical works
  - HVAC
  - Plumbing
  - Fire fighting
- Material handling (for solids handling plants)
  - Bunkers, cyclones, chutes, ducts, filters
  - Conveyors, elevators, rotary feeders

The disciplines that are assigned to one or more of the above duties may take names such as "vessels engineering," "heat transfer (or heat exchanger) engineering," "static equipment engineering," and "machinery (or rotating equipment) engineering". Alternate cases, like putting together in one discipline the vessels, heat transfer, and material selection under a name while assigning the rest of the mechanical duties to another discipline also exists in engineering companies. Depending on the type and size of the engineering company, all these duties may be performed by a single discipline, named as the "mechanical engineering."

The services involved in equipment engineering basically serve to specifying the equipment items required for the project. Vessel engineers review, calculate, draft outline drawings and standard details for tanks, drums, reactors, columns, etc., in order to develop the necessary requisitions for collecting bids from manufacturers. These technical inputs to manufacturers are to be selected with care by the engineer not to fix all manufacturing parameters, since certain design features have to be left to the manufacturer for applying its own proprietary methodologies and standards. Heat transfer engineers perform thermal rating of heat exchangers, but usually leave mechanical rating to the manufacturers. In any case, all further information or details calculated and designed by manufacturers have to be continuously checked and commented as required by the engineer, both at the quotation stage and later, during preparation of vendor data and drawings after order placement. The same principles apply to all groups of equipment and material handling systems. For all these equipment items, the shop drawings are always prepared by the manufacturers. So, one of the main tasks of the engineers participating in such activities becomes reviewing and commenting to vendor calculations, drawings, and other related documentation until correct and complete deliverables are received from the vendors.

Rotating equipment differs from static equipment in the sense that the manufacturers have certain models and types already designed and tested for a range of duties and conditions. Equipment including this type is unlikely available as already assembled for immediate delivery upon order. Performance characteristics of such equipment are often catalogued for review by purchasers. Hence, upon receipt of an inquiry, vendors select from their range of production and quote for the model that best suits the requirements. Then, the duty of the rotating equipment engineer is to check the technical part of the proposal for consistency and compliance with the requirements; for instance, for a pump, reviews are made for the vendor data consisting of capacity, differential pressure, range of operating conditions, NPSH (Net Positive Suction Head), efficiency and motor power curves as well as information about material of construction, seals, flushing, etc. Since such equipment items are fully proprietary designs of vendors, it is not possible to entirely verify by way of calculating back the design and to ensure that the vendor committed performance is correct until the item is performance tested. It is to be noted that process gas compressors, gas turbines, and steam turbines are much more complicated than pumps and so require extensive efforts during preparation of requisitions, reviewing of bids, and checking vendor drawings and information.

Material engineering activities include preparation of MOC drawings that actually start with mark-ups made on flow sheets at early stages of a project. These show the type of the construction materials specified for major plant components in accordance with the properties of processed substances and operating conditions. Similarly, the piping material specification of a project is prepared by the material engineers who then follow up and resolve material issues identified later in the project, including welding and testing.

Mechanical engineering for buildings requires dealing mainly with HVAC, plumbing and firefighting as well as other mechanical items, if any, like lifts, bridge cranes, etc. Pharmaceutical, synthetic fiber and some food plants mostly have very stringent HVAC requirements. Systems and equipment needed for these plants, like air handling units and fans are either ordered separately through requisitioning or left to construction contractors' supply. Plumbing design is required for potable/sanitary water distribution and sanitary discharges of shower and toilet areas. Firefighting systems inside buildings are usually treated in the same way, in compliance with the applicable codes and local regulations. All these inside-building distributions require preparation of MTOs. Meanwhile, engineering for the main firefighting system of the plant consisting of fire water pumps, water storage, plant fire water loop, hydrants, monitors, foam systems, etc., is performed either by mechanical discipline or by project engineers with support from mechanical discipline. Such engineering studies are basically performed for preparing the corresponding requisitions. For the final design configuration of the firefighting system as well as the hydraulic checks, recommendations by firefighting vendors are reviewed and incorporated in the design.

An outline of the duties performed by the discipline(s) responsible for engineering of heat transfer, vessels and material selection is as follows:

- Heat exchanger thermal rating
- Heat exchanger mechanical rating (usually as checking vendor design)
- Vessels engineering drawings
- Equipment platforms
- Heat exchanger requisitions
- Vessel requisitions
- Technical bid evaluation and vendor information review
- MOC drawings
- Piping material specification

The outline of the duties performed by the discipline handling rotating equipment and other miscellaneous mechanical works is as follows:

- Rotating equipment requisitions
- HVAC drawings, requisitions
- Plumbing drawings
- Firefighting design and requisitions
- Material handling design and requisitions
- MTOs for the above systems
- Technical bid evaluation and vendor information review

A summary of the main engineering activities performed by mechanical/equipment discipline(s) is given in Figure 8.11 in a simplified work flow diagram format.

The work flow diagram in Figure 8.11 is prepared by omitting interdisciplinary information exchange for the sake of keeping the figure simple. Technical bid evaluations and vendor information reviews also omitted are separately shown in Figure 8.14.

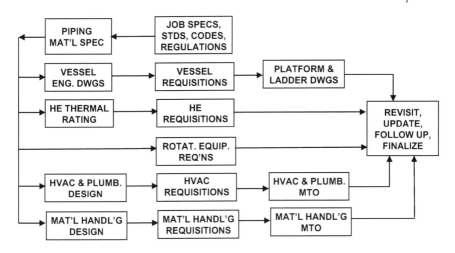

FIGURE 8.11 Simplified work flow diagram for major activities by mechanical/equipment discipline(s).

CIVIL AND STRUCTURAL ENGINEERING

Civil engineering is one of the oldest engineering disciplines in human history and deals with calculating and designing of buildings, structures, roads/paving, foundations, bridges, ports, grading, drainage, etc. It is the application of physics and mathematics together with principles of geotechnical engineering, surveying and structural engineering. Structural engineering is a specialty engineering discipline that deals with calculation, analysis and design of structures that support static and dynamic loads from any direction. It is generally considered as a sub-discipline or a section of civil engineering. Static loads, also called dead loads, include self-weight of the structure and the load exerted by an item on that structure. Dynamic or live loads cover forces applied by rotating/moving/vibrating objects (e.g., machinery), or by wind, seismic, or thermal actions.

It should be noted that architecture has often been used in conjunction with civil engineering. In old times, there was no clear distinction between civil engineering and architecture and so, either of these terms was used to refer to the same expertise. Later, in modern ages, architecture emerged to be a separate occupation from civil engineering and both have become more defined and formalized professions.

For industrial plants, civil and the structural engineering has a wide range of application. Design of plant sections and structures does not generally prioritize attaining architectural aesthetics, excluding buildings for administration, social buildings and gate houses inside plant sites. On the other hand, although many types of process units are located outdoors due to various operational, safety and cost reasons, certain types of production plants like, pharmaceutical and food plants are always enclosed in buildings mainly for hygienic reasons. In any case, architectural design takes usually its role in the work produced by this discipline due to the need for preparing building layouts, sections, and views whenever a building is included in the scope of the project. Accordingly, in the next paragraphs, the term "civil" will be used to include also architectural type of work, when such specialty is also involved.

Civil engineering is one of the disciplines performing its specialty functions in a project by exchanging engineering information both within the interdisciplinary environment as well as with the external parties. The external parties are involved for specific information to be provided by specialized external contractors in the beginning of the project under guidance of the civil discipline. These are mainly the geotechnical investigation report and surveying study report (topographical mapping) of the field. The geotechnical report provides the results of the analysis based on borings made at site with particular care to sample from locations that will be under heavy equipment or structure loads. The report provides very critical information for the civil engineering work, like soil bearing force, ground water level, any requirements for ground reinforcement (e.g., piling) and type of foundations recommended. Topographical map is prepared by surveyors using special equipment, like theodolites, for measuring angular deviations, horizontal, vertical, and slope distances. It then graphically shows on a drawing the existing site conditions, such as elevations and coordinates, with reference points (like, existing fixed structures or buildings), and markers to guide the construction of the project facilities.

Civil engineering is usually started with main structures, like interconnecting pipe rack, as soon as piping scope layout is ready.

For solid processing plants and utility units inside buildings, the preliminary equipment layout within the building are to be available for starting the civil engineering, based on preliminarily sized equipment and loads. Obviously, earlier task for such buildings is to have also the architectural layouts prepared for all the elevations to be set in the design.

For steel structures, the usual way is to prepare only single lines unless the steel structure manufacturer requests a more detailed design like, general arrangements and connection details. However, in any case, the manufacturing or shop drawings are expected to be produced by the manufacturer. For concrete structures, formwork plans are drafted. Production of the corresponding rebar schedules are normally left to the construction contractors.

As soon as the plot plan becomes sufficiently developed and ready for the first issue (i.e., after comments from other involved parties are resolved and incorporated), site grading/site preparation plans showing excavation, land fill, retaining walls, etc. can be produced. Completion of these drawings normally requires sufficient progress of design for foundation locations and underground facilities (i.e., sewers, underground (U/G) piping, cable trenches, grounding, etc.).

Drawings for equipment and structure foundations are another civil engineering task. These are usually sketched, based on estimated loads and then revisited for any modification that may be required after actual loads are received from the vendors. Foundation location plans are prepared to show all the foundations in relation to each other with reference dimensions and coordinates of the site.

Design of drainage (sewer systems) is started when drains from process, utility, wastewater, and storm water are available. Concrete containments, like cable and pipe trenches and culverts require information about routings of pipes and cables. Designing the underground is a typical extensive multidiscipline integration work that requires involvement of almost all engineering disciplines, as not only drains, cables, pipes, and equipment locations are of concern, but also other items including instrument junction boxes, grounding roads, any road under-passes, paving, dikes, curbs, etc., are to be jointly considered. It should be noted that the underground facilities together with site grading are normally one of the first construction works carried out at the jobsite, though the full U/G design requires a good level of progress in the overall engineering. Hence, for tight scheduled or fast track projects in which notable schedule overlap of engineering with construction is to be implemented, certain U/G facilities are constructed either by later re-excavating and re-filling or by estimating the missing information with some contingency.

If 3D design is to be applied to civil works, steel structures, foundations, and even U/G can be designed in 3D, integrated with other main 3D design elements like piping, equipment, cable raceways, etc.

As to summarize, the main categories of civil and structural engineering deliverables are as follows:

- Job specifications (support to PE)
- Site grading plans and details
- Architectural layouts, views, and details
- Architectural schedules
- Foundation plans
- Foundation location plans
- Formwork plans and details
- Rebar schedules
- Steel structure single line diagrams
- Steel structure general arrangements and connection details
- U/G plans and details
- Fence, road, and paving details
- Civil engineering calculations

An outline of the main civil engineering activities is given in Figure 8.12 in a simplified work flow diagram format.

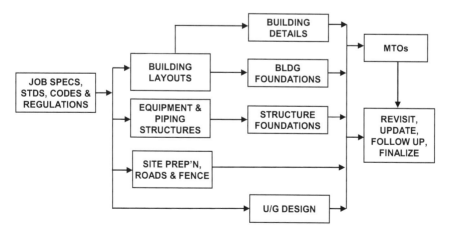

FIGURE 8.12   Simplified work flow diagram for major activities by civil engineering.

The work flow diagram in Figure 8.12 is prepared by omitting interdisciplinary information exchange for keeping the figure simple. Technical bid evaluations and vendor information reviews also omitted are separately shown in Figure 8.14.

ELECTRICAL ENGINEERING

Electrical engineering is the last engineering discipline described in this chapter. It practically starts a little later than the other disciplines and gets completed usually at the very end due to the fact that many of the necessary inputs to this discipline are developed by the other disciplines along the progress of engineering. These inputs are basically the plot plan, above ground and underground layouts, building drawings, equipment power requirements, etc. Such inputs provide design information to the electrical engineer about the locations and quantities/magnitudes/levels for feeding power, routing cables, providing lighting, designing the lightning protection/grounding as well as weak current systems basically for communication, and safety – security. During the period until receipt of all the necessary inputs, electrical engineers make their reviews, preliminary studies and comment to the design of the other disciplines.

One of the key electrical services needed to operate a plant is the power supply, its distribution and the accompanying grounding system. Many plants have incoming MV (medium voltage, 1–35 kV) or HV (high voltage, above 35 kV) power supplied from outside, i.e., from regional or national power grids. Some plants have power generation units inside and connecting additionally to the grid may still be an operational advantage. In such cases, a secondary power supply from the grid may be installed against worst black out cases of the plant power generation system, unless the facility is very remote from power grids.

Incoming power from outside passes through a substation in which the voltage level is reduced to the levels used in the plant (such as 3–6 kV for MV and 0.4 kV for LV) in transformers equipped with switchgear panels, both at the upstream and downstream to disconnect the circuit during downstream faults as well as during periods of maintenance.

In power plants, step up transformers are required to elevate the voltage level generated usually at around 10–14 kV to the level of the grid that the produced power is to be fed. If the power generation unit is used only for feeding the process plant, then voltage level is reduced to the levels required in the plant as described above.

In almost all plants, it is also necessary to install emergency power systems consisting of UPS and back up diesel generators in order to keep supplying power to critical loads until safely shutting down the plant on a black out.

Power distribution is furnished from power distribution panels by cables running from the distribution points to the users' connections either along cable trays/cable ladders/cable trenches or inside conduits. The main power distribution panels feeding motor consumers are named as motor control centers (MCC). These panels equipped with starters, relays, breakers, indicators, etc., are arranged in sections for each of electrical motors and other power consumers.

Once a preliminary load list (electrical consumer list) is made available, electrical engineers can start working on single line (also called, one line) diagrams that indicate schematically the entire power supply and distribution system. These diagrams represent key electrical drawings similar to the P&IDs for the process and are further developed during the progress of engineering as actual information becomes available.

Grounding of a plant is provided by a grounding network connected to grounding plates or rods, all buried in the ground to yield a selected resistivity. All grounding connections of electrical systems and metallic items like panels, steel structure, piping, etc., are attached to this network. Grounding layouts show how the system components are arranged on the site layout. Lightning protection is also a kind of grounding, but serving against lightning, instead of electrical system faults or static electricity build up. Conventional lightning protection is of two main types: Lightning rods and Faraday cage.

Lighting layouts and details are produced to show type and location of lighting fixtures for achieving the proper illumination levels, both indoors and outdoors. The type of fixtures is selected in accordance with the objective of the lighting (e.g., flood lights or projectors for general wide coverage, fluorescent, LED, or incandescent for local spot lighting like field instruments) as well as the area classification of the location. In a plant where flammable vapors or dusts may evolve, not only lighting fixtures, but also other electrical equipment (e.g., panels) and components (e.g., push buttons, cable connectors) to be placed in such particular areas are selected in accordance with the classified area.

Weak current systems operate at low voltage levels and consist of telecommunication (e.g., telephone, paging, loud speaker, fire alarm, security/surveillance, badging) and data systems. Some of the foregoing examples are engineered by instrument or even separate telecom disciplines in some engineering companies.

Routing cables along cable trays, ladders and conduits as well as supporting these require a sort of multidisciplinary review by instrument, electrical, piping and structural disciplines. It is a normal practice to leave sufficient margin on cable carrying elements for additional future cabling. Cable routing layouts, cable tray plans and sections are prepared, accordingly.

Requisitions for electrical equipment are prepared by electrical engineers. Such equipment includes transformers, switchgear, MCCs, other electrical panels, diesel generators, UPS, etc. The requisitions contain information on duties and requirements with attachments like, single line and schematic diagrams and job specifications. The requisitioned equipment items are engineered and designed in detail by the vendors for review by the electrical engineer both during technical bid evaluation and the reviews of vendor drawings received after order placement.

Electrical systems of plants require various types of components for connections, termed as electrical bulk material. These include cables, cable glands, conduits, conduit fittings, etc., all in different sizes and types. All these materials need to be counted and listed in cable lists and electrical bulk MTO. Supply of these materials is usually included in scope of construction contractors.

The main categories of electrical engineering and design documentation prepared for process plant projects are as follows:

- Job specifications (support to PE)
- Electrical load list
- Single line diagrams
- Substation layouts

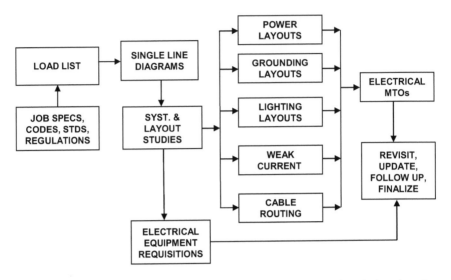

FIGURE 8.13 Simplified work flow diagram for major activities by electrical engineering.

- Power distribution plans
- Grounding layouts
- Lightning protection drawings
- Lighting layouts
- Weak current system layouts
- Electrical details
- Cable routing/Cable tray plans
- Electrical engineering calculations (e.g., short circuit analysis, relay settings)
- Cable lists
- Electrical equipment requisitions
- Electrical bulk MTO
- Technical bid evaluation and vendor information review

An outline of the major electrical engineering activities is given in Figure 8.13 in a simplified work flow diagram format.

The work flow diagram in Figure 8.13 is prepared by omitting interdisciplinary information exchange for the sake of keeping the figure simple. Technical bid evaluations and vendor information reviews also omitted are separately shown in Figure 8.14.

INTERFACE OF ENGINEERING WITH PROCUREMENT

The work flow diagrams given in above sections for activities of engineering disciplines do not show technical bid evaluations and vendor drawings—data reviews. As indicated before, this approach is selected for keeping the figures as simple and clear as possible. As a remedy, the relevant activities are separately indicated in Figure 8.14 as a chart which is common to all disciplines when dealing with technical information proposed by bidders and vendors.

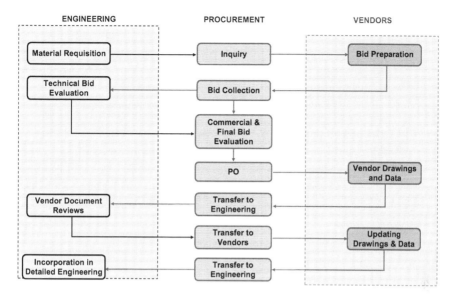

FIGURE 8.14   Simplified work flow diagram for interface of engineering with procurement activities.

PROJECT ORGANIZATION OF ENGINEERING DISCIPLINES

A typical organization scheme is shown in Figures 6.2 and 6.3 in Chapter 6 – Project Management, in form of an organization chart indicating the upper key levels taking part in a project. Each key level reporting to the project manager is further rolled down with necessary staff for producing the corresponding services. Engineering functions of a project team are usually organized under an engineering manager (or an engineering coordinator or sometimes, a project engineering coordinator) assigned to the project. Process engineering is often taken as a line separate from rest of the engineering disciplines as mentioned in Chapter 6 – Project Management.

Project organization of an engineering discipline is formed by way of assigning a lead engineer, sufficient number of discipline engineers, designers and drafters to the project by the discipline head. A typical organization scheme looks like as shown in Figure 8.15.

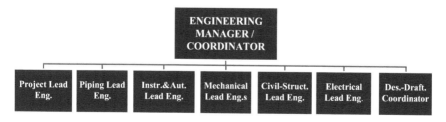

FIGURE 8.15   Project organization of engineering disciplines.

The above scheme may take some modified forms, depending on the size and scope of the project as well as the operating practices of the engineering company.

Lead engineers are basically responsible for the following duties:

- Ensure compliance with project requirements including the contractual scope, project specifications and procedures, applicable standards, codes and regulations and the approved time schedule.
- Share the work among the specialty engineers of the department and ensure that sufficient information is available at every stage of the work.
- Coordinate the flow of information to and from other engineering departments.
- Issue list of requisitions to be prepared by the discipline.
- Review requisitions prepared by the specialty engineers of the discipline.
- Review technical bid evaluations carried out by the specialty engineers of the discipline.
- Assist to construction contracting and purchasing departments.
- Check design calculations of special items.
- Ensure that all vendor drawings are reviewed for technical acceptability and compliance with requisitions.
- Ensure that design calculations by the specialty engineers of the discipline have been done based on latest updates of the project data and in accordance with the applicable codes and standards.
- Suggest corrective or improvement actions according to the quality assurance system.
- Coordinate and follow-up drafting activities of the discipline.

It is the lead engineer of each discipline who prepares together with the design-drafting coordinator the drawing lists, releases approved basic design data, releases certified vendors' drawings for incorporation in the design, ensures that information obtained from project engineers, other specialty engineers and vendors are transferred to the design-drafting coordinator and drafters.

VERIFICATION OF ENGINEERING AND DESIGN

All outputs from engineering and design activities need to be checked for correctness, consistency, and compliance with the project requirements. The main references for the verification are typically the following:

- Project execution documents (contract, PEP, Minutes of Meeting (MOM), project correspondence)
- Applicable codes, standards, and regulations
- Owner's standards/specifications
- Vendor/subcontractor drawings and data
- Quality system documentation

Project deliverables (i.e., calculations, data sheets, requisitions, drawings) including supplier and subcontractor documents are all to be checked before they are released for use in further steps of the engineering and in construction. The methods and responsibilities for checking, review, and approval of project deliverables are clearly defined in the engineering quality system documentation. These steps for each deliverable are marked on the corresponding deliverables in form of signatures by the persons authorized to perform each of the foregoing set of verification steps. Each set of such signatures on a document means that the indicated revision of that document is released with data and formulas used as well as the design shown on that document are consistent and correct in accordance with the project requirements. All revisions and the verifications to the project deliverables are handled by the discipline which performs the preparation, checking, and approval of the same.

Verification is performed continuously throughout the engineering and design activities including all design changes, which—though not desired—do inevitably happen in all projects. Verification techniques typically include the following:

- Checking and approval of design outputs prior to release
- Interdisciplinary circulation of documents and data for interdisciplinary verification and comments
- Formal design reviews (P&ID review, model review, etc.)
- Use of alternative calculations
- Comparison with previous similar designs

Verification of engineering and design is one of the most important steps in the entire chain of project implementation activities since building up the subsequent engineering and design of the project on consistent and correct grounds bear a crucial importance in the success of a project.

FORMAL DESIGN REVIEWS

Hypothetically, any engineering and design deliverable may be reviewed and commented for approval by the owner or its representatives (e.g., the Project Management Consultant, if exists in the project). Practically, only the key engineering—design documents, like P&IDs, single lines, plot plans, main layouts, major requisitions, etc., are selected by owners to be subject of review and approval due to time concerns. Additionally, any applicable regulatory institutions (approval authorities) may require making same type of reviews for granting certain permits, like the construction permit. These are regular reviews carried out during implementation of a project.

Formal design reviews are similar to the above with regard to the main objective, but differs in the fact that the latter are performed in dedicated sessions and attended by a group of persons including the design originators, other interfacing disciplines, owners and sometimes third parties, like facilitators, external experts, etc. These reviews are scheduled at the beginning of projects and performed at appropriate stages of engineering and design by properly documenting all comments for following up the subsequent incorporation. The formal design reviews are typically as follows:

- Basic design or FEED review
- P&ID review

- Plot plan review
- Model review
- Constructability review
- HSE reviews (HAZID, HAZOP, etc.)
- SIL review

DESIGN AND DRAFTING

In engineering companies, the term "design" is mostly used to refer to setting up configurations, layouts, views, and details, rather than engineering calculations.

Designers are experienced design experts possessing a technical degree in either one of mechanical, civil, or electrical subjects, not necessarily in engineering. It is mostly the designers who design layouts, select design details (and so, fabrication and construction details), make MTOs, prepare bulk material requisitions and solve design detail problems, like clashes and non-conformities.

Drafters are CAD operators drafting the design, by using software in 2D or 3D or both, in accordance with the design-drafting standards used in the project. Today, almost all drawings are produced by using CAD, which superseded the manual techniques applied until few decades ago. These previous techniques involved drafting by hand on special transparent paper by using either ink or special drafting pencil. The reason of using transparent paper was to enable making copies by blue printing, as today's photocopiers that can copy large size drawings were not available then. Drafters were using special drafting boards equipped with sliding rulers. Symbols and wording were marked on drawings either by free-hand or by using templates. Apparently, it was taking significantly longer time to draft and most important of all, it was obviously impossible to make major revisions on an already drafted drawing. So, drawings that require major revisions had to be discarded and re-drafting on a new blank sheet from the beginning. Thus, major design revisions were likely to bring in significant re-doings and causing notable loss of time and extra drafting man-hours.

Engineering companies have usually separate drafting groups for each engineering discipline, each with specific experience and competence for the discipline they are part of. Accordingly, piping, civil, instrument, electrical and each of mechanical departments have their own drafting staff.

DOCUMENT MANAGEMENT

Management of documents for a facility project requires a quite careful and continuous attention. Engineering documents are generated during the project by engineering companies, vendors, and subcontractors for exchanging and integrating design information during project development as well as for manufacturing and construction when drawings reach IFC status. The documents to be managed can be grouped basically as follows:

- Specifications
- Calculations

- Data sheets
- Requisitions
- Indices (e.g., equipment list, line list, instrument list, etc.)
- Drawings/diagrams/sketches
- Isometrics
- Lists of deliverables (e.g., specification list, requisition list, drawing list, etc.)

When any document is ready for issue, its distribution is made in accordance with a distribution matrix prepared in the beginning of the project. This matrix, printed on document transmittal forms, shows names of recipients together with the name of their discipline and company. The recipients are mainly the owner and the project team members of the engineering company. Owner receives such documents either for information or for approval, based on owner's options discussed and agreed during the contract negotiations or in the beginning of the project. The purpose of internal distribution within the engineering departments is for exchanging design information and receiving comments from other disciplines in order to provide proper integration of the multidiscipline design.

Each single document requires to have a proper and unique identification, clearly shown on the document by

- A document number given in line with the numbering system used in the project.
- A revision number (or letter) with the date of the revision.
- A title box showing the following:
  - Purpose of that revision (e.g., for information, for approval, for construction)
  - Signatures of the persons involved, shown as "prepared by," "checked by," and "approved by"

A document number is like a car plate number; it should identify a particular document. In other words, no two documents can carry the same number. A document number is usually composed of a couple of groups of numbers and letters. Each company has its own document numbering system that may be very much different from others. However, all mostly have, as the first group in the document number, the project code or number assigned to that particular project by that company. The other groups of numbers or letters may indicate the area/unit code, the type of job (e.g., structural steel, piping layout, electrical power) and most importantly, the sequence number to segregate documents having the same codes before the sequence number. For instance, if there are 20 piping layouts for a particular unit, then the sequence number starts from 001 and ends at 020.

For any change made in a document, the revision number should also be carried forward (e.g., if document in revision 1 is modified, it becomes revision 2).

The "purpose of issue" shown on the title box bears a specific significance. A document issued for owner's approval should be followed up whether it is approved or approved with comments or rejected during the contractual period allowed for the owner's review. The contractual period for approval is usually 2 or maximum

3 weeks. If approved, then that document can be used as a basis or reference during subsequent activities in design. For no response received within the contractual duration of time, it is then deemed as approved and treated similarly. However, if approved with major comments or rejected, the document should immediately be modified as required and re-issued in order not to get delayed in the project. Another significance of the purpose of issue is during construction. Drawings that do not carry "issued for construction" statement (and often accordingly stamped) are not allowed to be used at site.

Document lists (i.e., specification list, requisition list, drawing list, isometric list) are prepared and continuously updated to show the latest revisions of each such deliverable. These lists also bear their own revision numbers and dates. Each engineer and designer working on the project should ensure that he or she is working with the latest revision of a document, whether it belongs to his/her discipline or to other disciplines or a vendor.

Hence, document management is an important aspect of engineering and design that requires due care. If not properly handled, costly mistakes and time taking re-doings would be unavoidable. However, it is quite easy to properly conduct document control by uploading all revisions of documents, including the ones by vendors, timely in a relatively simple computer program that can then show the latest status of all documents and is accessible to all authorized persons. Such software also makes the distribution to the persons in the distribution matrix. All the foregoing document recording and distribution activities are handled by a particular dedicated staff, often named as document management or document control group.

ENGINEERING STANDARDS, CODES, AND REGULATIONS

Standards, codes, and regulations are developed, updated, and imposed for application by regulatory agencies, engineering societies, trade and manufacturing organizations, and other state and private institutions. They all serve to ensure the quality, reliability, efficiency, and safety of equipment, systems, materials, structures and processes. Although each of these three terms has a distinct characteristic meaning, they are often used informally (and incorrectly) as if they are synonyms.

Engineering standards ensure that organizations and companies adhere to accepted professional practices (also termed as good engineering practices), including design, manufacturing, and construction techniques, personnel safety, environmental protection, etc. Standards can be regarded as proposed guidelines based on criteria and techniques developed in time and updated through accumulated experience and lessons learnt. A technical standard is an established norm or requirement and is usually a formal document that establishes uniform engineering or technical criteria, methods, processes and practices in order to ensure quality, reliability, and safety.

A code consists of set of rules and specifications for the correct methods and materials to be used in certain products, services, activities, buildings, and/or processes. Codes are often accepted by governments to set out compulsory requirements and, so carry the force of law. The main purpose is to protect the public by setting up the minimum acceptable level of safety for equipment, systems, materials, structures, and processes.

Engineering regulations are government defined practices enforced for ensuring in general the protection of the public. Manufacturing regulations usually involve legislation for controlling the practices of manufacturers that affect the environment, public health, or safety of workers.

Almost all countries have their standards, codes, and regulations, which sometimes refer to other such international documentation. Some examples to institutes/societies that develop and update internationally accepted standards and codes are as follows:

- ANSI (American National Standards Institute)
- ASTM (American Society for Testing Materials)
- API (American Petroleum Institute)
- NFPA (National Fire Protection Association)
- BSI (British Standards Institute)
- EN (European Standards maintained by CEN, CENELEC, and ETSI)
- ISO (International Organization for Standardization)
- DIN (Deutsches Institut für Normung)
- NEC (National Electric Code)
- ASME (American Society of Mechanical Engineers)

It is essential to list and agree with the owner, at the start of a project, on the standards to be used throughout the project. Applicable codes and regulations of the country/region of concern are compulsory to follow.

ENGINEERING SOFTWARE

Today, most of the engineering activities are carried out by using engineering software, though it was only some decades ago that the introduction of electronic calculator performing engineering functions was regarded as a revolution that immediately replaced the slide rule. This was followed by programmable calculators. During that time, computers were available, but were big clumsy main frame systems that required user to write the software and punch in cards for both the software and the data. Introduction of personal computers (PCs) and their further development in terms of capacity, speed, and user friendly operating systems happened to come out much later. Meanwhile, huge advances on the side of software have been materialized. Significant drop in prices of hardware made the personal computers affordable and enabled extensive use at work and at homes. Similar developments have also been accomplished for software. As of today, there are many computer applications that make the previously difficult calculations possible and easy. This also covers complicated simulations for which such modern software usually has inside a wide library of data. It is apparently important that use of such complicated software requires certain knowledge, training, and experience in properly using the software.

Some of the engineering activities that most commonly make use of computer applications are as follows:

- Process simulation and calculations
- Instrument sizing

- Piping stress analysis
- Civil/structural calculations
- Cable sizing
- Electrical short circuit analysis
- Tank and pressure vessel mechanical calculations
- Heat exchanger thermal and mechanical ratings
- Document management
- 2D and 3D drafting including intelligent systems starting from P&IDs

Many computer programs are available in the market for analysis of material, components, systems, emissions, etc., and new ones are continuously released for use by engineers.

9 Project Control

Project control is a function that directly supports project execution by focusing primarily on the project progress and project costs in order to make evaluations for the management. The principal objective is to continuously check the current status of a project, make projections for end of the project and evaluate them with reference to the baselines (i.e., the project schedule, project budget, and scope). In this respect, project control resembles indicators on a car's dashboard including a navigator that show the driver current location, recommended routes, the estimated time to reach the destination, the speed, the fuel level, any malfunctions, etc. Despite having the word, "control" in the name of this function, the real control is under the command of the driver, who is in reality the project manager (PM).

In Chapter 6 – Project Management – the duties and responsibilities of a PM are stated to cover "planning, organizing, monitoring and controlling" of the main project parameters that are namely the "time, cost, resources, scope and quality." In process plant projects, the overall "planning" and "monitoring" tasks for "time," "cost," and "resources" are delegated to project control for handling by way of data collection, evaluation, and reporting. The intent is to timely identify any significant deviations and adverse trends that would require taking immediate corrective actions in the project.

Successful completion of a project highly depends on effective use of good project control practices applied from the very beginning to encompass all stages of the project (i.e., engineering, procurement, construction contracting, construction, commissioning, and startup). This basically requires the following:

- A clear project scope and a quality plan are available to be the basis for the schedule and cost baselines.
- Project execution strategy and risk analysis are already reviewed and concluded.
- A realistic and adequately comprehensive project schedule is prepared in the very beginning.
- An approved realistic budget is available for implementation.
- A proper change management procedure exists.
- Procedures and techniques are available for collecting project control data including KPIs (Key Performance Indicators).
- Procedures and techniques are available for generating results and making analysis.
- Effective communication and document distribution channels are already in place for receiving and sharing information.
- A proper project organization is established.
- There is experience available in the project organization for performing objective and reliable project control functions.

From project control perspective, the above points represent building blocks of a sound project plan and are used to establish the project baselines. Project scope, budget, and schedule are basically the major items of project baselines for which continuous monitoring, data collection, analysis, and reporting are performed. Other KPIs, such as the ones related to quality (e.g., number of Non-Conformity Reports (NCRs)) and Health, Safety, and Environment (HSE) (e.g., number of near-misses) are also monitored as these are also to be controlled for project success. In that sense, a baseline is a benchmark against which all future project control activities are to be performed during the entire project execution stages. Baselines are often impacted by approved project changes, so they should be adjusted pursuant to change control activities in the project and updated, accordingly.

All these efforts are for the major objectives of predicting the following:

- Where the project is now and how far it is from where it is supposed to be according to the baseline
- Where the project would end and what deviations would get materialized in line with the current trends
- How to avoid (or at least, to minimize) negative deviations basically in completion time and expenditures as well as other undesired developments in the project

Obviously, a successful project is the one that gets completed on time and within budget by fulfilling all contractual and legitimate requirements without compromising on quality and HSE. Accordingly, one of the conditions for ending up with a successful project is to have a properly functioning project control.

In quite a number of engineering and construction companies, project control function is set up as a department which is not directly connected to project management or project execution division. This is for ensuring a neutral, independent, and realistic approach by this function all along the project. However, obviously, planners—cost controllers are assigned to work in project teams, but still under follow-up of the project control department.

A more common approach developed lately in project control is to have every project controller to deal with both the planning and cost control of projects, instead of having separate specialists for planning-scheduling and cost control.

PLANNING AND SCHEDULING

Planning is one of the most important techniques used in management of both profit and non-profit organizations. It is basically the process of making objective and realistic forecasts for reaching to the destination point within a predetermined future time through a cluster of consecutive and interconnected activities. Each of these activities requires a careful analysis for viability, resource requirements, anticipated duration, and dependency to other activities. Once a plan is completed and agreed/approved by the project stakeholders, it will then become the baseline and require monitoring and updating as necessary by adding latest forecasts on the existing actual status. Good planning-scheduling increases efficiency and reduces time of completion, resources to be employed and expenditures to be made while minimizing risks involved in the contract.

In process plant projects, planning is one of the major responsibilities of project management, performed by project control staff by way of preparing various schedules, however, under review and guidance of the PM. Project plans other than schedules, like the project execution plan, quality plan, and HSE plan are prepared either by the PM or by the relevant specialists, but always reviewed and approved by the PM and often the owner before becoming formal. For a project, the first implementation plans are prepared by project owners at conceptual stages. When a decision to kickoff is given, prospective contractors are requested to prepare and offer with the bids their execution plans, often with reference to owner's preset milestones. Such plans cover a proposed project schedule, other supporting plans and procedures, describing how the project will be performed if awarded to their company.

For every process plant project, a particular main schedule is composed to become the formal time plan (baseline) upon approval by the owner. It covers entire stages of a project from initiation of engineering up to end of commissioning and startup. Such schedules are also definitive references for schedules prepared by others, e.g., subcontractors and vendors.

Project schedules are timetables, prepared mostly in graphical form. They show planned activities and milestones as a function of time in a chronological sequence determined according to logical dependencies of activities with each other. At start of preparing a project schedule, the scheduler requires to know in detail the project scope and stages as well as the project obligations on timing. This is actually the list of activities and milestones, prepared in a structured approach, termed as work breakdown structure (WBS). Next is to pick the tools (i.e., scheduling technique and software). The contract of a project is normally the main source of information for the scope and timing requirements. It may also set out owner's requests for schedule levels and the tools to be used.

WORK BREAKDOWN STRUCTURE

WBS is a hierarchical arrangement of project work scope and services in form of levels that break down to bring in more details. The details can be rolled up to summary levels when required. It shows all phases, work packages, and fundamental deliverables of the project. In that sense, it provides a systematic structure for identifying and expanding by dividing into increments the deliverables and work packages. WBS is normally arranged for main products or outcomes, rather than actions that are also focused in projects. Anyhow, WBS is a key working document that divides the scope into manageable parts or packages that the project team can easily focus on. It helps defining project tasks within each activity to be used in developing the schedule. Additionally, it helps cost accounting by eliminating risk of duplication since WBS is based on the outcomes. It is a common approach to number (or code) WBS elements sequentially for providing the hierarchical structure, by which each category is further split down into elements with sequential numbers starting with the code of that category.

A typical approach in developing a WBS is to start with the highest (top) level, which is the final deliverable, e.g., the project itself. WBS is just a comprehensive classification of project scope, showing what will be done. Obviously, it is neither a project plan nor a schedule.

SCHEDULE LEVELS

Schedule levels are identifications about the level of details covered in a given schedule. These are mostly designated by way of numbering the levels. There is also an alternate way, called the descriptive identification, according to which the schedules are named as project master schedule, summary master schedule, detailed schedule, look-ahead schedule, etc.

No universal agreement exists for the numbering, as well as for the contents of schedule levels. However, it is a common practice in this industry to have five levels starting from level 1 which represents a general overview showing only major project activities, key deliverables, and milestones. As more details are incorporated for each item of a given level, the level number increases from level 1 up to level 5. Some professionals also mention about level 0 which is though not practically a project control level as it represents the total project, mostly with a single bar indicating only project start, duration, and finish.

A brief description of the schedule levels is as follows:

Level 1: This is a summary level of schedule highlighting major project activities and milestones, showing usually one bar for each of engineering, procurement, construction, and commissioning by major work packages and/or areas. It is often prepared during initial feasibility studies and also for supporting proposals. It may also be formed by rolling up (summarizing) level 2 schedules for reporting to senior executives.

Level 2: It is developed by further breaking down the items in level 1 or formed as a summary of level 3 for project execution. It is normally used for higher-level management reporting, addressed to general managers and project managers.

Level 3: It is prepared in logic network form by using Critical Path Method (CPM) method or as a summary from level 4. This type of schedule defines the critical path(s) of the entire project and is a primary control tool for project managers, construction managers, and owners' representatives. It is usually prepared by contractors for submitting with their bid and used for developing schedules with higher details (higher levels).

Level 4: This level is prepared by expanding level 3 and contains detailed information for each function and work package. It is a CPM-based working level schedule showing the activities to be performed by the project staff and the labor. Care should be taken to keep it in a reasonable size for ease of application and updating. Level 4 schedule may be prepared for only certain major portions or functions (e.g., construction, commissioning) of the project. These schedules are monitored by project managers, functional managers, discipline heads, and leads.

Level 5: Obviously, this level is developed by further breaking down the level 4 schedule. The detailed activities have much shorter durations for day-to-day follow-up and control of the project tasks in specific areas. It addresses project team members including supervisors, team leaders, superintendents, and foremen. It is often prepared to further detail certain portions of the project schedule, such as look-ahead and disciplinary schedules.

As mentioned above, different parties involved in a project have all different levels of interest and details needed in the project schedule. Owners are most likely interested in key milestones and start and completion dates at higher (summary) levels. Contractors require much more detailed schedules for monitoring and controlling their progress effectively in the work committed in accordance with the contract. The level and form of schedules to be submitted to the owner by a contractor is mostly dictated by the terms and conditions of the contract.

Schedules are initially prepared in a top-down approach and later, maintained in a bottom-up method, e.g., by rolling up from level 4 to lower-level numbers. The schedule structure should be set up as aligned with the WBS of the project in order to form similar basis with all other project control activities.

The optimum size of a process plant project schedule is a controversial matter; however, a widely accepted conception for a level 4 schedule is to have no more than one or maximum a few thousand activities for ease of understanding, follow-up, updating, and printing.

SCHEDULING TECHNIQUES

Various techniques are in use for planning, analyzing, and reporting project status with respect to time. Some of these are very simple tables indicating key milestones or certain critical tasks and showing the corresponding due dates. These are normally used only for reporting or follow-up purposes. On the other hand, there are more comprehensive methodologies used for detailed planning, monitoring, and control. Such more elaborate techniques for scheduling are as follows:

- Gantt (bar) chart
- Program Evaluation and Review Technique (PERT)
- CPM

Gantt Chart

Gantt chart is actually a bar chart, devised by Henry Gantt in the beginning of 20th century. It has a tabular form in which the first column lists the activities and the subsequent columns represent the time scale in days, weeks, or months, below which bars are placed corresponding to each activity shown in the first column. Hence, the length of each bar shows the duration of that activity, with its left edge showing the starting time and the right edge, the completion time as shown in Figure 9.1.

This technique can be used alone for uncomplicated projects having a limited number of activities as well as for level 1 or level 2 schedule of any other project, provided that the planner is experienced enough to incorporate correctly the durations of all the activities as well as their interdependencies. Due the simplicity of this graphical format, logic networks (i.e., PERT and CPM defined in the next paragraphs) are converted to and printed in bar chart forms after the results are obtained by processing the inputs on a computer.

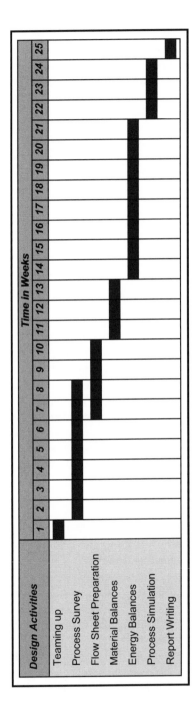

FIGURE 9.1   A simple bar chart.

Today, Gantt chart is still the most common scheduling format used in projects due to its simple look which makes it very easy to understand and interpret.

Modern Gantt charts are often arranged to show also dependency of each activity with others, as well as percent completion of activities by marking the current time as a vertical line across the schedule chart (the "time now" line). During project execution, each activity is represented by two sets of bars, one for the plan and the other for the actual progress achieved up to the "time now" line; then, the bar is continued to show the future forecast if the activity extends beyond "time now."

Program Evaluation and Review Technique

PERT was developed for the US Navy around the end of 1950s to support Polaris Nuclear Submarine Development Project as a statistical technique for measuring and forecasting progress. This method requires the following inputs:

- List of activities
- Expected duration of each activity – PERT uses three sets of time data: (p) Pessimistic, (o) optimistic, and (m) most likely time duration; the expected time is then calculated as $(p + 4m + o)/6$
- A logic network diagram showing connections of each activity to one or more of other activities according to its interdependence as predecessor and successor

Figure 9.2 shows a simple logic network on "Activity on Arrow (AOA)" format with the activities designated as A to J on arrows, while the numbered circles are the nodes showing starting and ending points. This format has in time been largely replaced with "Activity on Node (AON)" format described in the next technique. Once the network is developed with other necessary information made ready (i.e., relations of each activity with others and activity durations), time calculations can be performed as forward pass and backward pass to determine the critical path. The calculation method, though quite simple, is not going to be explained here, as they are outside the scope of this book.

Critical path is the longest route of the activities within the network in terms of time. It determines the time of completion of the project. Any delay in one or more of these critical activities postpone the project completion time as much as the delay in the critical activity. Calculations are performed with input data processed on a

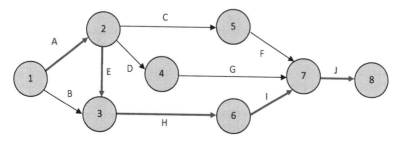

FIGURE 9.2 A simple activity network drawn in AOA format.

computer. If network consists of less than few dozens of activities, manual calculation is also possible, though still not very practical. Calculations yield the total project duration as well as a set of results for each activity, such as the time for Early Start (ES), Early Finish (EF), Late Start (LS), Late Finish (LF), and slack (i.e., the float indicating the period of time that a noncritical activity can be delayed before it becomes a critical activity) and, of course, the critical path.

In fact, PERT method is normally recommended for use in mainly the R&D (Research & Development) projects in which time estimates have significant uncertainties, so that a probabilistic approach by using three sets of time data is applied. However, such a feature of this technique makes it somewhat impractical to apply to engineering-construction projects that possess up to a thousand or more activities. This method finds much less application than its rival, the CPM, in many areas of application.

Critical Path Method

CPM is created by Du Pont and Remington Rand in late 1950s. This is about the same time with the development of PERT.

CPM possesses very similar features with PERT with respect to the following:

- Project activities are identified with predecessors and successors.
- A similar logic network diagram is prepared based on predecessors and successor of each activity.
- Data is processed on a computer, using an appropriate software.
- Calculation is performed as forward pass followed by backward pass.
- The output yields the critical path of the network of activities and the total project duration.
- The results also include ES (Early Start), EF (Early Finish), LS (Late Start). LF (Late Finish) and slack (termed also as float or total float) for each activity,
- The network configuration can be printed with calculated data.
- The schedule is printed in form of a bar chart for ease of review and interpretation of the results.
- They are both used to analyze in detail the completion time of projects.

Major differences between CPM and PERT are as follows:

- PERT is basically a statistical method focusing events that have significant uncertainties with regard to times of completion. It can provide some statistical data such as probabilities and deviations.
- PERT uses three sets of time data (optimistic, pessimistic, and most likely), while CPM uses only one set for activity durations, mostly estimated through historical data that is deemed reasonably accurate. That is why CPM is referred to as a deterministic method.
- CPM can also focus easily on resources and costs of activities.
- CPM is flexible and practical enough with regard to its preparation, modification, and updating in facility projects, and it is the most commonly used tool for such purposes.

Despite the fact that the original CPM program form (i.e., in AOA method) is no longer used, the same name commonly refers to the network logic technique modified in time from the original version to AON form for a more practical application. A sample network logic diagram, drafted in the AON format, is shown in Figure 9.3:

In the AON format, the activities are placed in nodes, shaped as boxes, instead of putting them on arrows as in AOA format. Within sections of each box, relevant time information is shown. In AON, arrows are used to show the relations of activities with each other. This format is also called precedence diagramming method.

An activity box usually appears as indicated in Figure 9.4.

The advantages of this new form of CPM are basically as follows:

- Ease of showing activity information clearly.
- Eliminating the dummy activities, which do not correspond to any real activity, but frequently required in AOA format. Dummy activities have zero time of completion and are used only to indicate start–end relations of arrows, when required. The number of such non-real activities may come out to be as many as real activities; thus, unnecessarily complicating the network.
- Providing more flexibility by using activity relations other than only FS (Finish to Start: A successor may only start if its predecessors are finished). The aforementioned additional relationships include FF (Finish to Finish), SS (Start to Start), SF (Start to Finish), Start with Lead, Start with Lag, Finish with Lag, etc.

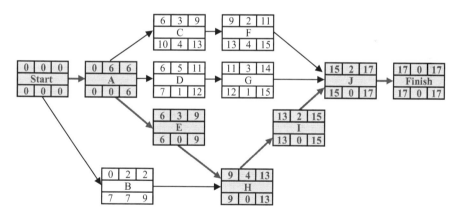

FIGURE 9.3 A simple activity network, drafted in AON format.

ES	Duration	EF
Name of the Activity		
LS	Float	LF

FIGURE 9.4 Form of an activity node in AON format of logic networks.

- The resulting schedule can be printed in form of a bar chart in different schedule levels by rolling up and down. Interdependence of the bars, like FS, FF, SS, etc., can also be shown on the bar chart as vertical arrows.

FURTHER WORK ON SCHEDULES

Once the schedule is prepared in required detail and approved for implementation, it is updated and issued periodically (usually monthly, but sometimes also biweekly or weekly according to project requirements) with reference to actual progress of the works. While developing the schedule, optimization of the resources should be considered for better management of the resource requirements, minding corresponding effects on cost and time. Such techniques include the following:

- Resource leveling: In this technique, start and finish dates of activities are adjusted in line with resource constraints, basically for balancing demand for resources with the available supply and may result in change of project duration.
- Resource smoothing: This technique is applied after resource leveling and refers to further adjustment of activities for achieving that the resource requirements do not exceed certain desired limits. This method can be applied to activities which have floats and does not change the duration of the project.

During execution of the project, planners may need to apply modifications and include further details to the forthcoming steps of the scheduled work for improving, controlling, and remedying the project forecasts in terms of time, cost, and resource allocation. Some of such modification methodologies used for schedule recovery or acceleration, with care not to increase risks above tolerable levels, are as follows:

- Crashing: Speeding up certain activities mainly on the critical path by adding more resources
- Fast tracking: Performing some of the activities in parallel for acceleration of the schedule without allocating additional resources

PROGRESS REPORTS

For process plant projects, it is essential to have periodical reports, often monthly, prepared for highlighting current project status with regard to the progress achieved in the project up to that month. Monthly reports—also called monthly progress reports – mainly consist of the following:

- A narrative section, describing and comparing the following with the plan:
  - Cumulative actual physical progress achieved in the past month, as overall and work category level, e.g., engineering, procurement, and construction
  - Activities performed by each function for all the project stages under execution

- – Actual status with regard to the time schedule
- – HSE and security statistics
- – Summary of key issues, mitigations, and achievements
- – Any areas of concern
- Progress schedule tables of major work categories, as well as more detailed lower levels, like each engineering discipline, each subcategory of construction works, etc., showing planned and corresponding actual physical progress values as a function of time in monthly intervals.
- S-curves and manpower histograms for engineering, procurement, fabrication, construction, and commissioning.
- Level-1, level-2, and level-3 bar chart schedules extracted from the latest update of the logic network.
- Verification of productivities with predictions on progress trends.
- Photographs showing progress status at the construction site.
- Plans for the next month and look-ahead schedules.
- Discussions and conclusions for outstanding developments and areas of concern.

Narrative sections of progress reports are prepared by the PM, after analyzing the latest studies performed by the project control function, such as updated schedules, analysis of trends, S-curves and histograms. The entire monthly report is issued by the PM.

Physical Progress

Physical progress term as used in analysis and reporting of project status stands for the best realistic value of the actual progress in a particular work item. For instance, completing half of the paving of site roads corresponds to 50% physical progress. It is different from man-hour progress as the work might have been accomplished by spending 60% of the man-hours planned. Accordingly, physical progress refers to real completion level of the project outcomes. Physical progress values of all project activities are then processed to estimate the physical progress of the entire project and are also used to verify productivities.

In engineering and design, physical progress is based mainly on deliverables, e.g., calculations, specifications, data sheets, drawings, and design details. Analysis is started from the bottom level by estimating physical progress of each deliverable using applicable milestones, like completion of drafting, checking, first issue of the drawing, receipt of owner's approval, and making the for-construction issue. Then, with the bottom-up approach, physical progress on the disciplinary level and then the overall functional or phase level are calculated.

A similar approach is followed by engineering companies for procurement services, the physical progress of which is tied up to issue of deliverables, such as procurement procedure, vendor list, inquiries, bid tabs, purchase orders, and procurement reports.

It is more straightforward for construction which is based on actual quantities of work done for each particular work item, such as volume of concrete poured. Similarly, progress of commissioning is based on sub-systems and systems commissioned.

COST CONTROL

Cost control basically starts with establishing the cost baseline prepared from the approved budget in connection to scope and time baseline. As the project progresses, cost control carries out monitoring the cost performance, analyzing the findings, forecasting for completion, and comparing these with the baseline. If results indicate any significant negative deviations or adverse trends, the situation will need to be rectified through sound adjustments and necessary precautions. Diagnosis for deviations and adverse trends is only possible by continuously repeating the "monitoring-analyzing- forecasting-comparing" cycle all along the project execution.

Main baseline for cost is primarily the approved budget (baseline control budget) which is further split into smaller elements and arranged in form of code of accounts (COAs) for ease of follow-up and analysis. Cost breakdown structure should be compliant with the WBS. Project cash flow plan is also an important duty of cost control function, prepared as a time-phased plan of project expenditures.

CODE OF ACCOUNTS

A project COAs is a system of numbering expected project costs by setting up main cost categories and then splitting those down to further details. The coding is usually made up of numbers or letters or a combination of both. Each company, whether it is an owner or a contractor, has usually its own system for COA.

For a process plant project, the main category (top level) items for COAs may be set up as follows:

- Home Office (HO) services (project management, engineering, procurement, and other direct HO services)
- Direct materials (equipment, system packages, and bulk material),
- Construction contracts/subcontracts (construction works and third-party site inspection costs)
- Field indirect services (construction supervision, commissioning, and vendor supervision)
- Field facilities (office equipment and supplies, temporary facilities, and utilities)
- Other expenses (insurance costs, bank charges, performance bonds, training, etc.)

Each of the above categories is then further split into one or more levels of details as needed. For instance, the category of direct material often requires to be further split into the following:

- Static equipment to cover vessels, heat exchangers, etc.
- Rotating equipment consisting of pumps, compressors, turbines, etc.
- Packaged systems including utilities, firefighting systems, miscellaneous material handling systems, etc.

- Structural steel (if not included in construction costs under the scope of a subcontractor. Some portions of structural steel may be contained in scope of package vendors.)
- Piping material including valves, pipes, fittings, flanges, gaskets, and stud-bolts
- Instrumentation and instrument bulk materials (bulks are usually included in construction subcontractor's scope)
- Electrical equipment and bulks (bulks are usually included in construction subcontractor's scope)
- Protective covers including insulation, noise abatement, etc. (if not included in scope of construction subcontractor. Some portions of protective covers may be in the scope of package vendors.)

The third level of COA below the second, if needed, is established based on a further distinctive feature. For instance, rotating equipment at level 2 can be further split down to indicate the sub-type of rotating equipment as centrifugal pumps, gear pumps, reciprocating compressors, etc. It may alternately show directly the equipment tag numbers, like P-512 A/B or K-202, for all items involved at this level. A short-cut alternate is to skip entirely the third level and insert all those additional details under the category titles of the second level.

Experienced and well-structured companies in process plant industry use their standardized COA, both during estimating and project execution. The most important point in setting up the COA is to ensure that no cost item is either omitted or counted twice.

It is very important that the codes of account are defined since the pre-award phase, i.e., during the proposal. In this way, the same structure can be adopted easily from the cost estimating files to be inserted in the cost control system files.

Cost Data

Project cost data to be used by project control primarily consists of the following:

- Committed costs: Costs corresponding to COA items already ordered or contracted, so liability for payment is undertaken.
- Estimate to complete: Costs estimated to be accrued and paid in the future on top of already committed items.

For proper performance of cost control, timely flow of information and documentation on all purchase orders placed, contracts awarded, invoices received and all other expenditures should be provided to cost controllers. Information about man-hour charges to the project (through timesheets) with corresponding man-hour rates, expenses by project staff (through expense reports) and indirect expenses should be made available as well. Hence, cost controllers require working hand in hand with procurement, subcontracting, and accounting departments.

Cost Reporting

For process plant projects, cost control outcomes are analyzed and reported in form of periodical (often monthly) cost reports. These reports are built up from bottom level details and rolled up to higher levels, ending up at the summary level as of the date of the report. The summary page is often an abstract for the analysis results, showing also the latest contract price that may have been altered due to possible change orders. It often lists the results from the previous periodical reports in a tabular form. Values of general overhead, profit and reserves (for completion and make good) can be indicated and analyzed at this summary level.

The pages beneath the summary are mostly in a tabular form showing COA starting from top level and rolling down to lower levels. For each COA item, the original budget, latest approved budget, committed payments, total forecast and budget variance values are given.

A sample form of such a table for top level COA is shown in Table 9.1.

Table 9.2 shows a sample table for a second level of COA, selected to indicate further split of HO services. Other second-level categories, as well as the third-level classes, if any, are all similar in having identical columns for cost data, but obviously with different COAs on the rows in the first two columns.

The objective of cost control and reporting is apparently to monitor closely and take controlling actions when adverse trends in expenditures against control budget are diagnosed. Budget overruns may arise from changes developed or requested by the owner at any time during the progress of the project execution. For such changes, a change management system is to be in place from the very beginning of the project so that change identification, processing, and owner's approval steps are specified and implemented as required. Approved changes are added to the budgets and so, the updated current budgets are formed, accordingly.

TABLE 9.1
A Sample Cost Table for Top Level of COA

COA	Category	Original Budget ($M)	Approved Changes ($M)	Current Budget ($M)	Committed to Date ($M)	Estim. to Complete ($M)	Total Forecast ($M)	Budget Variance ($M)
1000	HO Services	3,900	135	4,035	3,780	350	4,130	−95
2000	Direct Materials	42,690	975	43,665	40,414	1,475	41,889	1,776
3000	Direct Subcontracts	11,700	1,035	12,735	10,935	2,550	13,485	−750
4000	Field Supervision	2,135	72	2,207	2,123	212	2,335	−128
5000	Field Facilities	235	0	235	211	0	211	24
6000	Other Expenses	540	0	540	420	85	505	35
0000	Total	61,200	2,217	63,417	57,883	4,672	62,555	862

TABLE 9.2

A Sample Cost Table for Second Level of COA (Example Is for HO Services)

Category	Original Budget ($M)	Approved Changes ($M)	Current Budget ($M)	Committed to Date ($M)	Estim. to Complete ($M)	Total Forecast ($M)	Budget Variance ($M)
1001 Project Management	400	15	415	375	70	445	−30
1100 Engineering Services	1,600	110	1,710	1,388	200	1,588	122
1200 Procurement Services	350	8	358	300	40	340	18
1300 Construction Contracting	90	0	90	90	5	95	−5
1400 External Services	1,060	2	1,062	1,050	30	1,080	−18
1900 Indirect Expenses	400	0	400	400	5	405	−5
1000 Subtotal	3,900	135	4,035	3,603	350	3,953	82

TIME RECORDING AND MANAGEMENT

Man-hours expended by the project personnel generate part of the project costs in the course of performing the services (i.e., project management, engineering, procurement, construction management, etc.). For lump-sum projects, such as the ones on Engineering-Procurement-Construction (EPC) basis, man-hour costs of a contractor are to be recorded and managed in order to control that the expenditures plus the future forecast are in line with the corresponding budgets. For man-hour reimbursable projects, such as the ones on Engineering-Procurement-Construction management (EPCm) basis, man-hours spent by the contractor are invoiced to the owner at the contractual rates. In such man-hour reimbursable projects, man-hour costs are also to be monitored and controlled with respect to the owner-approved budgets for avoiding payment disputes. These requirements necessitate putting in place a time recording and management system, regardless of the contract basis. Such a system is to be applied in line with a dedicated procedure and based on filling in timesheets for recording the man-hours that the project staff spends for the project. Timesheets are filled in periodically (often weekly or sometimes daily) and processed to sum up the charges by each manpower category from the start of the contract. Time management system provides an important set of cost data to the cost control function of the project.

10 Procurement

Procurement is another important function which should be started during early stages of project implementation. It should go hand in hand with the progress in engineering in order to have the supplies delivered before they are required for construction at the jobsite. Accordingly, as the engineering makes ready the specifications of equipment, and bulk materials, preparation, and issue of inquiries are commenced for the purchasing activities.

Procurement can be reviewed under the following three sub-categories that complement each other as consecutive stages:

- Purchasing
- Expediting and inspection
- Logistics/transportation

At the start of a project, the procurement activities are basically related to the purchasing phase and initially consist of getting prepared by setting up the following:

- Procurement strategy
- Procurement procedure

PROCUREMENT STRATEGY

Similar to other project strategies, the procurement strategy is the base plan, tailor made to incorporate the project requirements and the market conditions relevant to supply of equipment and materials.

The procurement strategy mainly covers the following:

- Basis of inquiries (competitive bidding or direct negotiation)
- Markets to supply from (local and/or acceptable foreign markets, compulsory markets due to financing terms)
- Bid acceptance priorities (lowest price or shortest delivery time among acceptable bids)
- Delivery (the way to handle freight forwarding, transportation, cargo insurance, custom clearance)
- Methods for expediting (phone calls, e-mails, visits to vendors' premises)
- Attendance to tests and inspection (no participation, selective participation, contracting out to third parties)

Inquiries are mostly prepared for competitive bidding unless the owner has some other intents or obligations relevant to some of the supplies. Such special cases may arise from the following:

- Owner's long-term agreement with any of the potential suppliers
- Intention to minimize spare part inventories by maintaining common spares with other units/plants of the owner
- Correctly interfacing the new system with an existing one in terms of hardware and software (typically the case for DCS)

Selecting the markets from which the supply is to be made basically depends on the following:

- Local availability of the supplies
- Terms of the financing that may enforce buying from the country of the financer
- Quality versus price considerations

Normally, orders are placed to the technically acceptable lowest priced bids. However, in very tight-scheduled projects and especially for the long delivery items, this might not be a good way to follow. Then, the successful bid becomes the technically acceptable lowest priced bid having an acceptable delivery time.

Setting up the delivery terms involves determining the supplier's responsibility with regard to transportation of the supply. Sometimes, the supply is received from the vendor's facility as ex-works and all subsequent transportation arrangements are handled by the purchaser with or without using a freight forwarding agent and a custom broker.

Vendors visits for expediting and inspection is another decision to be taken and included in the procurement strategy. Making visits for all or only equipment items or only for selected critical items are to be clearly defined.

PROCUREMENT PROCEDURE

Procurement procedure is prepared in the beginning of a project in line with the procurement strategy. It is a detailed description of the techniques to be used in the procurement activities of the project. It defines how to deal with each step of the procurement as well as what formats and reports to be used for these services in the project. More descriptively, it covers the following:

- Main highlights from the procurement strategy to be applied
- Preparation and approval of project vendors list
- Preparation and issue of inquiry packages
- Obtaining quotations from bidders
- Bid openings
- Technical and commercial evaluation of the bids
- Selecting the successful bidder
- Preparation of letter of intent and formal purchase order

- Order revisions
- Vendor drawings and information
- Expediting, inspection, and release for shipment
- Logistics/transportation
- Material handling at jobsite
- Invoice checking and payment approvals
- Formats, templates, and standard documentation to be used all along the procurement activities (e.g., Request for Quotation (RFQ) form, instructions to bidders, general purchasing conditions, bid tabulation forms, letter of intent format, purchase order format, procurement reports, etc.)

Purchasing may sound to be a very simple and straightforward task to persons who have not been involved in process plant projects. However, anyone with such experience knows how critical it is and so, how systematically this task should be carried out for avoiding disastrous situations. It is very critical because supply of acceptable material from reliable suppliers with suitable delivery times and at prices within budget is one of the most important targets for the success of a project. The items supplied are the real tangible outcomes of extensive engineering and management that can be achieved as desired only through proper purchasing activities. It has to be very systematically performed in order to

- Avoid incomplete or incorrect orders.
- Eliminate future disputes.
- Control the process not to cause any delivery delays.
- Receive all necessary approvals from management (and the owner when contractually required) for avoiding any difficult corrective actions and negotiations after order placement.
- Form strong, reliable contracts with suppliers through purchase orders.
- Ensure receipt of correct and reliable material conforming to the requirements.
- Minimize impacts of damages and losses during transportation.
- Document every step for any future queries, discussions, and any legal actions.
- Provide reporting and filing as evidence for compliance with quality management as well as code of ethics.

PURCHASING

As soon as the procurement procedure is set up and accepted by the management as well as the owner, the following purchasing steps can be performed:

- Preparation of vendor list
- Preparation of the inquiry documentation
- Inquiry issue and clarification of bidders' commercial and technical queries
- Receipt of quotations from bidders
- Technical and commercial bid evaluations

- Selection of the successful bidder
- Preparation and issue letter of intent or directly the purchase order
- Follow-up and timely receipt of vendor drawings and information
- Checking and approving vendor invoices
- Arrangement of freight forwarding and custom clearance
- Timely delivery of the supplies

Vendor List

This is actually a "list of acceptable/qualified bidders," prepared to name material, equipment and service suppliers and/or manufacturers that are deemed to be suitable for the project. Selection criteria usually depend on the following:

- Reliability and reputability (based on either direct experience or indirect knowledge of performance in engineering; sub-supplying; manufacturing; health, safety, and environment; quality management; delivery as well as financial status)
- Nature of the project (for projects with complicated and stringent requirements, bidders that can work to such standards and provide all the required certificates are to be selected)
- Size and nature of the supply (e.g., for relatively small-sized orders, material suppliers may be inquired rather than manufacturers)
- Brands already used elsewhere in the plant or in some other plants of the owner (for reducing the inventory size of spares or for problem-free incorporation of the item into an existing system)
- Conditions of project contract and/or finance agreement (that may set forth nominated vendors and/or countries to supply from)

For any new vendor to be included in the list, a pre-qualification through a questionnaire and then a facility visit is often made as a condition to check the adequacy of the vendor prior to its inclusion in the list. As soon as the vendor list is prepared and approved by the management, it becomes one of the key documents for purchasing. Any modifications (i.e., additions or deletions) may only be incorporated in the list upon consent of the owner. Other requirements related to the vendor list are usually as follows:

- Project vendor list is to state for each bidder, its official name, category of materials, equipment and services, the country, and the contact information.
- The list is usually to cover minimum three (if available) and maximum seven or eight bidders for each category.
- In general, inquiries are issued to a minimum of three bidders in order to have a reasonably good update about the current market prices as well as to maximize the probability that the best price and delivery terms are obtained. If less than three valid responses are received, management should have the option of either soliciting quotations from other bidders or

to select a successful bidder from quotations in hand. Single source materials and services should be highlighted in the vendor list.

- More than seven or eight bidders increase the cost of purchasing services and the time required to condition the bids.

INQUIRIES

An inquiry package is a set of documents that describe the requirements of a supply, both technically and commercially, while giving instructions to bidders on how to bid. Inquiries are either sent to or given access in web to the bidders that are in the project vendor list. An inquiry package usually consists of the following documents:

- RFQ
- Instructions to bidders
- General purchasing conditions
- Requisition and attached data, specifications, and drawings

RFQ is a form letter addressed to each bidder separately, indicating the inquiry name, inquiry number, and contents of the package as well as the bid closing date.

Instruction to bidders explains the bidding procedure, the required bid information and the format of the bid as well as the contact information of the purchaser.

General purchasing conditions set forth the applicable general liabilities of both the seller and the purchaser for a possible order placement subsequent to bidding. This is basically a legal contract between the purchaser and the seller to become effective upon placement of an order. Sometimes the vendors ask applying their own terms and conditions instead of the one received with the inquiry. Then, it becomes a matter of negotiation during the bid evaluation stage and may lead to disqualification of the bidder. It is important that general terms and conditions should be reasonable, fair, clear, and not open-ended; otherwise obtaining good bids may be at risk.

The requisition (sometimes called the material requisition abbreviated as MR) is the technical part of the entire package that describes mainly the scope and technical requirements of the supply including the technical documentation to be produced by the vendor. For equipment and package systems, necessary inputs for process and engineering information are presented in form of data sheets, job specifications, and drawings that are needed for preparing the bid. Descriptions are made particularly to clarify the requirements including the type, size/capacity, configuration, material of construction, and the applicable standards and codes as well as the desired performance characteristics of the supply. Special instructions are also furnished regarding the requirements for vendor drawings and data.

INQUIRY ISSUE

Before issuing an inquiry, it is a good practice to contact the listed bidders and ask them about their interest to bid. Inquiries can then be issued in line with the procurement schedule, with more assurance that the planned number of bids will be obtained. In any case, instruction to bidders normally asks the bidders to acknowledge receipt of the package and intention to quote. In general, the quotation time allowed is around 2 to 4 weeks, depending on the size and complexity of the supply.

Bidders receive inquiry packages mostly via courier or sometimes by accessing purchaser's FTP server on the web through internet with the password e-mailed to the listed bidders.

While preparing quotations, bidders may need some technical and commercial clarifications. In such cases, they have to make those in writing to the contact person of the purchaser. It is the duty of purchaser to respond to each such query. If the question is concluded to be essential or helpful to the bidder, then the contact person of the buyer arranges a clear answer by consulting, as required, other specialists in the purchaser's organization. If not, the answer may state "bid as you see it." Any clarifications made to a bidder, should also be sent to all other bidders in the inquiry.

It is not uncommon that one or more bidders request extension to the bid closing date during the inquiry stage. In such a situation, it is usually the PM who reviews the following:

- The project schedule, especially the critical paths
- The number of bidders remaining in the inquiry if extension request is refused
- The risks of not receiving good or complete bids from the quoting bidders as well as the possibility of other declines to bidding and decides, accordingly

Prior to bidding on specially engineered equipment, or complicated system packages, pre-bid meeting(s) may be held to review fully and clarify any further the design requirements.

RECEIPT AND OPENING OF BIDS

Bidders must submit quotations by the stated bid closing date. Late bids are not accepted unless specifically approved by the project management.

Bidders are usually required to submit the following:

- Priced original bid for commercial evaluation
- Unpriced bid copies for technical evaluation

Bidders are often asked to submit quotations in sealed envelopes as the contents are quite critical and sensitive for carrying out a fair and professional purchasing campaign. Sealed bid envelopes are kept in a secure cabinet until the opening date. Alternately, bidders may be asked to load their bid on the web to an FTP server by using their passwords assigned for that inquiry.

Bids are normally opened in a closed session one day after the bid closing date by a committee composed of minimum two or three persons assigned by the purchaser (the engineering contractor that prepared the inquiry package). The owner is also represented in the committee unless the purchaser's contract with the owner is on Lump-Sum Turn-Key (LSTK) basis.

During the bid opening session, major bid information, like scope of supply, prices, delivery, and other important commercial terms offered by each bidder, is recorded in a tabular form, usually named as preliminary bid summary at opening, or bid opening summary. This table mainly serves to

- Give rough information about the initial outcome of the inquiry, before bid conditioning.
- Represent an initial record along the development path of the inquiry, basically for any future investigations against any possible allegations.

Immediately after the initial recording of the quotation information, unpriced bid copies are forwarded to engineering specialist for technical evaluation. The priced original and copies are kept by the authorized purchasing staff for the commercial evaluation. These are handled only by authorized personnel in order to guarantee that the required confidentiality is met.

Once sealed envelopes are opened, any bid subsequently received should not normally be taken into consideration during the next steps of the inquiry handling activities. For inquiries with an estimated value below a couple of thousand USD, it may be decided to omit sealed bid procedure. In such a case, quotations received by fax or e-mail are acceptable.

Any information or result of bidding or any developments in an inquiry, whether technical or commercial, should never be leaked to anyone not authorized for dealing with the inquiry.

Bid Evaluation

The first step in bid evaluation is to have the preliminary bid tabulation completed through a systematic review of the bids as received. This enables pre-screening – usually on a price level and delivery time basis – of the bidders whose quotations are to be further evaluated in detail. In other words, such an overview through preliminary bid tabulation allows screening out the quotations which are very far from being acceptable due to very high price, very long delivery time and/or significantly incorrect or incomplete supply. This in turn will help saving time in the subsequent detailed evaluation.

A maximum number of four to five bidders that quoted reasonable prices, acceptable delivery times and are generally in conformity with the inquiry can then be identified as the short-listed bidders for detailed technical and commercial analysis.

During detailed analysis, evaluators identify necessary questions and clarifications to be asked each bidder. Such requests are conveyed to the bidders by the nominated contact person in charge of the inquiry. Obviously, e-mailing questions and receiving the corresponding answers in full takes significant time and so, it is unlikely possible to finish conditioning of an inquiry in less than 2 weeks. This takes much more time for bids covering complicated and quite a number of items.

Once all necessary clarifications are obtained, the final technical evaluation of the bids can be completed. In case of problems in receiving full satisfactory responses, technical clarification meetings may be held with the bidders prior to issuing the final technical evaluation. Alternately, bidders that fail to provide timely the necessary clarifications may be disqualified by the consent of the project management (and the owner in reimbursable projects) in order to avoid running out of time.

Simultaneously with the technical evaluation, bids are also commercially reviewed in a similar manner. At the end of this stage, an "after technical bid tabulation" can be prepared to report vendor(s) recommended to be called for the final negotiation.

Final negotiation meetings may be held, when necessary, with those bidders that rank at the top in terms of price and delivery in addition to full technical acceptability. Such meeting target ensuring the scope of work is precisely understood, all commercial conditions are fully accepted and the final negotiated price is agreed. For all these meetings with bidders, minutes are written by the purchaser and forwarded to the attending bidder for acceptance.

When all the necessary inquiry conditioning activities, including final meetings if any, are completed, the final bid tabulation that indicates final prices and the main commercial terms is prepared and issued for approval of the management. This tab clearly states which bidder is to get the order, as being the best bidder of the inquiry. After approval of the contractor's management, similar consent needs to be received from the owner in reimbursable projects.

Order Placement

Material, equipment, and services are normally purchased from the "technically acceptable lowest priced bidder," which means "the bidder with the total net first low cost after adjustment for the required scope of supply, offering an acceptable delivery time and committing compliance with the negotiated and agreed purchasing conditions."

Accordingly, any award to a bidder

- Who submitted late quotation
- Whose offer is higher than the lowest bid
- When there are less than two competitive quotations
- When competitive bidding is waived due to urgency

normally requires consent of the management (and the owner in reimbursable projects) for any such waiving.

Upon approval by management of final bid tabulation, a "letter of intent" may be prepared, signed and transmitted to the selected vendor. Letter of intent normally indicates briefly the scope and price of the supply, delivery terms, and the most important clauses relevant to the upcoming formal purchase order. Letter of intent is to be used if such an approach is deemed to save time; otherwise, this intermediate step may be skipped and a purchase order may directly be prepared and issued.

Formal purchase order is to reflect all conditions of the procurement agreed by the parties involved. As an attachment to the order, it is also necessary to update the "requisition for order" by way of modifying the previous "requisition for inquiry" to reflect the final agreed conditions with the successful bidder. Purchase orders should clearly specify lump-sum prices, unit prices, discounts, total cost, terms of payment, delivery date, shipping terms, and instructions. Additionally, all orders should describe inspection requirements, guarantees and warranty terms, and documentation required for customs clearance.

After placement of orders, it is a good practice to ask vendors' immediate acknowledgment of order receipts by e-mail. Expediting order acknowledgments from vendors may be of crucial importance for a project since an order acknowledgment not received may mean the order not properly formalized with a possibility that the vendor may unexpectedly ask for deviations or exceptions. Anyway, it is the best to identify any such inconvenience as early as possible and implement immediately necessary actions for final settlement.

Warranty period is mostly set as 12 months from start-up or 18 months from delivery whichever comes earlier. For orders over a predefined monetary level, vendors are usually required to provide irrevocable and unconditional bank guarantee letters corresponding to some portion of the total order value (e.g., 10%). These bank guarantee letters are kept at hand for any possible liquidated damages for performance and released when the warranty period expires with no performance issues within that period. It is also a common approach to include in purchase orders certain back charge conditions for any delivery delays, usually termed as "liquidated damages for delay." Liquidated damages for delay are usually set as 0.5%–1.0% of the order value for every week of delay in delivery of the materials, with a cap of around 5% of the order value. The conditions regarding liquidated damages for performance and delay should be presented to the bidders in the general terms and conditions of the inquiry package from the beginning of the inquiry stage.

For itemized supplies (tagged items), like equipment, instruments, etc., it is also necessary to purchase the spare parts. Commissioning spares and special tools for the supply should actually be considered as a part of and delivered together with the main supply. Operational and capital spares may be ordered later, through a new order dedicated to such spare parts. However, in any case quotations for all the spares should be obtained with the main inquiry for the purpose of extending competition also to the spares and so, avoiding possible high quotations that may come for the spare parts after placing the order for the main items.

Below is a simplified work flow diagram (Figure 10.1) for the inquiry stage showing main purchasing activities by the purchaser for each inquiry.

ORDER FOLLOW-UP

After order placement, the next stage is the order follow-up. It continues actively until end of the delivery. Even if the supply is received without any defects, damages,

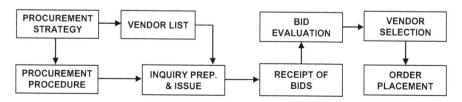

FIGURE 10.1 Simplified work flow diagram for purchasing activities by purchaser during the inquiry stage.

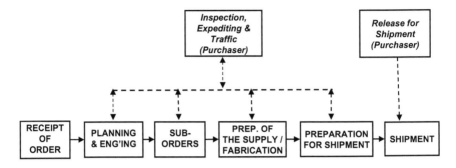

FIGURE 10.2 Simplified work flow diagram for vendor activities following a purchase order.

or shortfalls at site, the follow-up is still not closed out until end of warranty period. During this stage, the following activities are performed:

- Verify receipt of order acknowledgment and bank letters by the vendor.
- Keep the contact with the vendor during vendor's processing of the order.
- Arrange inspection, expediting, and logistics activities.
- Issue procurement reports and follow up compliance with procurement schedule.
- Follow up receipt of vendor drawings.
- Check vendor invoices.
- Follow up receipt of material at site and have the delivery checked for compliance with the requisition as well as for its condition at receipt regarding any possible damages during transportation.
- Verify performance of the vendor with regard to any developments that might have occurred for application of liquidated damages for delay, liquidated damages for performance, and/or any other back charges.

A simplified work flow diagram for vendor activities after order placement is presented in Figure 10.2.

INSPECTION AND EXPEDITING

Inspection is a planned examination or formal evaluation exercise and involves document reviews, visual checks, instrumental measurements, cold or hot functional checks, and other type of tests (like radiographic, magnetic particle, and/or dye penetrant examinations), applied to parts or to the entire body of a material item, equipment, instrument, or system. The results are compared to pre-specified requirements and standards for determining whether the item is in line with the specifications. Inspections are almost always non-destructive.

It is the vendors' responsibility to carry out the inspection programs dictated by the terms of the order which gives direct references to the relevant standards to be applied as well as to the inspection specifications that may be attached to the orders.

Although the responsibility lies with the vendor, for the success of the project, the purchaser normally follows up and often participates by witnessing these inspection sessions. The inspection documentation and the material certificates then become an important record for that item all along its operational life.

Expediting is one of the main activities in purchasing for securing timely delivery of material, equipment, and components in strict compliance with the order requirements and the contractually committed quality. Thus, expediting targets ensuring the supply is competently prepared by the vendor in line with the schedule, so that it is delivered at the promised time, at the agreed location and exactly as specified in the order. Expediting bears special importance in the following:

- Medium- to large-scale projects in which cost of delays are very high
- Any project in which the schedule is quite tight

Expediting is performed in accordance with expediting plans prepared for each order. This type of work consists of reviewing the progress achieved by vendors through phone or e-mail communications and expediting visits to vendors' shops. Each step in supplying/manufacturing the order is verified if properly arranged or already done in line with the schedule, including the engineering, sub-supplier orders and deliveries, manufacturing and testing. Each expediting call or visit is followed by an expediting report. Similarly, each inspection visit is to be documented in an inspection report.

The expediting and inspection activities consist mainly of the following:

- Preparation and issue of the expediting and inspection plan for each order
- Performance of home office expediting activities via phone, email, etc.
- Coordination and scheduling expediting/inspection visits to vendors' premises, including pre-production and pre-inspection meetings
- Ensuring the implementation of expediting/inspection plan for each order
- Review and analysis of vendor fabrication schedules
- Review and approval of vendor QC System, inspection data books and material test certificates
- Checking progress of vendor's engineering and sub-suppliers' delivery
- Following up fabrication progress
- Witnessing shop testing
- Checking compliance of the materials and equipment with the order requirements
- Preparation of expediting/inspection reports after each shop visit
- Expediting corrective actions to resolve production and delivery problems
- Final Inspection and release of materials for delivery

The "inspection and expediting activities" are mostly performed by engineering contractors who are the actual buyers on LSTK projects or act on behalf the owner on reimbursable contracts. Accordingly, these engineering contractors normally possess such disciplines teamed up with inspectors and expediters. An alternate approach can be to subcontract those activities to third parties.

LOGISTICS

The main duty of the logistics section of the procurement department is to manage qualification and selection of freight forwarding and receiving agents as well as selection of the carriers. Logistics activities also include managing or ensuring to have arranged the cargo insurance and custom clearance for the supply. All these activities are mostly handled through an external freight forwarder and custom clearance agent assigned for the project.

Logistics activities basically include the following:

- Issue release for shipment notes to vendors and suppliers.
- Ensure that correct shipping documents are made available by vendors.
- Coordinate the shipping agents and custom clearance agent (or freight forwarder doing both).
- Monitor the receipt and storage of the materials and equipment.
- Follow up customs clearance operation.
- Coordinate inland transportation.

For additional details on logistics, transportation, and customs, please refer to Appendix 8 – Transportation, Custom Clearance, Incoterms.

PROCUREMENT REPORTS

During procurement activities of a project, a number of reports are prepared and issued to management and the owner for providing information on progress as well as areas of concern. These reports present monitoring information and also represent records on purchasing for possible audits or investigations. These are mainly as follows:

- Inquiry status summary
- Bid tabulation
- Order list
- Material progress report
- Expediting/inspection reports

Inquiry status summary is prepared periodically and it contains information about the entire inquiry stage of the project by listing all inquiries with bid closing dates and the status of each inquiry as of the date of the report.

Commercial bid tabs are prepared at different stages of each inquiry and named by the stage that the tab is prepared for (i.e., "preliminary at opening," "preliminary," "after technical tabulation," and "final"). Backed up with technical tabulations, these records document all the important information and developments about each inquiry, giving bidders' names, prices, delivery terms, payment terms, guarantees, and other terms and conditions as well as the final recommendation by the purchasing group at completion of the bid conditioning.

Another periodical report is the order list which documents key information on the orders placed until the date of the report, including the name of selected vendors, material descriptions, corresponding requisitions for order, order dates, order amounts, etc.

Material progress report outlines the status of procurement for each requisition periodically, highlighting if the material is inquired, ordered, ready for shipment or on shipment (if so, the expected delivery dates also indicated in the report).

Expediting reports are those reports prepared by expeditors of the purchaser after each expediting session, such as a visit to vendor's premises. The main objective of such reports is to verify if proper and timely actions are being taken by the vendor to meet the promised delivery date. Inspection reports are prepared by inspectors of the purchaser for each test made by the vendor and witnessed by the inspector in accordance with the order. For tests not attended by the purchaser, the vendor prepares and submits the test report for approval.

RECEIPT OF MATERIAL AT JOBSITE

One of the last steps of each order is taken at the time when the order is received on a vehicle at the jobsite. Even when all prior inspection and expediting activities are performed by the purchaser including formal checks and release for shipment, it is still important to check the supply upon arrival at site. This will serve to minimize the impact of possible delivery problems listed below:

- Vendor might have transported incorrect cargo that belongs to some other buyer or to some other project of the purchaser.
- There may be missing packages or components, either lost on the way or not loaded at the vendor's premises.
- There may be shortage for some of components or bulk items supplied with the order.
- The supply or some components of the supply might have been damaged during the transportation.

Accordingly, it is essential to properly check the materials as soon as they arrive at site in order to take any necessary actions promptly without losing time. Below is a basic checklist for the checks:

- Check that the driver hands over all the related shipping documents.
- Check through the consignment note that the goods received are for the project.
- Conduct a package count and check for any visible damages to packaging and/or goods prior to off-loading.
- In case any damage or loss is noticed, then the delivery note should be signed indicating the damage or loss.
- Delivered items should be identified against related purchase order and requisition and should be checked for compliance with supplier's packing list.
- If the materials received are subject to overage, shortage, or damage, then the first action to be immediately taken is preparation of the corresponding discrepancy report.

- In any case, the carrier's delivery note should be signed as "received – unexamined" for any defects or inconvenience that may be detected later.
- After above checks, if there is still any doubt on the correct or undamaged receipt of the material, then inspection should be performed whenever possible and practical. Such materials are to be stored in a separate location (sometimes called as quarantine) at the warehouse or lay down area and kept untouched/unused until cancellation of the discrepancy report.

11 Construction Contracting

A project, kicked off for establishing an industrial facility progresses along a number of stages and all these are just for the basic goal of constructing the desired plant in the end. Accordingly, when prerequisite activities, basically in engineering and procurement are sufficiently developed for starting the construction works at site, it is then time to commence with contracting (or subcontracting) these works.

"Contracting" is obviously not performed only in selecting and making agreements with construction companies. This term is used also for selecting any other contractor, like consultants, engineering companies, third-party inspectors, etc., as well as non-technical services, like external trainers for team-building sessions or catering services for the temporary site facilities. However, the main focus in this section will be on the "construction contracting."

The term "contracting" changes to "subcontracting" when the main contractor is serving the owner on Engineering-Procurement-Construction (EPC) or Engineering-Procurement-Construction management (EPCm) basis by bearing the construction responsibilities against the owner, contracts out the construction directly with such responsibilities born by itself. Construction contractors may also subcontract part of their scope of the work.

Construction contracting is very similar to purchasing with respect to the techniques, concepts, and procedure followed, especially in preparing list of bidders, inquiry packages as well as bid collections, openings, evaluations, and awarding the contracts.

Construction contracting can be split into two main stages:

- Contracting stage
- Contract administration stage

The first stage above is a home-office function for the activities starting from preparation of contracting strategy, contractors list and construction inquiries up to and including preparation of construction contracts. The second stage is mainly the administration and following up of the awarded construction contracts at the jobsite for essentially resolving possible contractual issues, like claims, extra work requests, and disputes.

At the start of a project, the early construction contracting activities consist of setting up the following:

- Construction strategy
- Construction contracting procedure

CONSTRUCTION STRATEGY

Construction strategy is the major plan for setting out how to carry out the construction works of the project. It is developed as the outcome of the evaluations leading to selection of the options among alternatives.

It should be noted that the only way for implementing process plant projects is not to start with engineering and to proceed with procurement by either using an engineering contractor on EPC, EPCm, or EP (Engineering-Procurement) basis, or alternately, having these activities performed by the owner's staff. Another way of approach can as well be to select a PMC first, either for only management of the project for and on behalf of the owner and sometimes for also doing the conceptual engineering and/or FEED and then for preparing EPC packages to be awarded to other contractors for the rest of the project services. Such an approach is a frequently used option in large and complex projects, mainly for refineries and petrochemicals. The following descriptions in this chapter apply to both cases, which are either the construction subcontracting by an EPC/EPCm contractor or construction contracting by an owner. Both cases will be referred to as construction contracting.

For the construction phase, it is always recommended to clearly analyze the project with regard to its nature, size, schedule as well as prevailing market conditions before deciding on the number of construction contracts for the project. When the size of the construction works is notably large, as in case of multiple-unit projects, it may be advisable to employ more than one construction contract, even for similar trades of work, say mechanical works. The latter will bring in the flexibility in terms of ability of transferring work from one to the other when there are performance problems of contractors. Splitting the work into a number of construction contracts, including relatively minor work items such as temporary facilities, site grading, structural steel, heavy lifts, painting, and insulation, may be considered depending on the size of such categories of work. Apparently, either way, has certain pros and cons. Splitting into multiple contracts may provide a number of advantages, such as the following:

- Early start and orderly progress of site works in line with the progress of engineering and material supply (e.g., engineering for site grading can be made ready for construction much earlier than most of mechanical and electrical works. Additionally, site grading is the first of the site works to be performed during construction.).
- Employing contractors that are specialized in each specific category of the work.
- Saving cost from the markup percentages that a single main contractor would otherwise add to the price of the other contractors when using them as its subcontractors.

On the other hand, the same approach may lead to the following disadvantages:

- Managing more contractors requires more supervision and increased supervision cost.
- Difficult to resolve interference problems may develop among the contractors.
- Each contractor will require separate temporary site facilities which may create site allocation problems.

If the size of the construction work is rather small, a single construction inquiry may be more appropriate for attracting interest of good construction companies due to allocating the entire construction contract budget to a single contract. This is because capabilities and experience level of medium to large size contractors are normally higher than the small ones and such companies often have a minimum contract value for jobs to quote. Another advantage of a single construction contract is elimination of interfaces and possible disputes that may arise due to multiple contractors at the same site. However, for a single construction contract, the contractor will have to anyway subcontract part of the work that is not in its own field of expertise, though under its responsibility and control.

The contracting strategy should set up also the basis of construction contracts as lump sum, unit price, or other. The selected pricing basis is a major parameter to be applied throughout the inquiry documentation. For instance, material take offs or bill of quantities are never to be provided in a lump-sum inquiry for eliminating possible later claims by contractors. However, a unit price contract is based on estimated quantities which may in reality, increase or decrease to certain extent and thus directly vary the price. Setting up the pricing basis actually depends on the progress of engineering achieved at the time of contracting. Mostly, when progress of engineering is close to 80–90%, lump-sum contracting is regarded as the best choice and gives the advantage of relatively faster progress of the work with the expectation of an almost fixed total price. Extra works leading to cost increases are never avoidable in any circumstances, but are normally much less when progress of engineering is high at the time of construction contracting. This apparently requires also that all other conditions are properly arranged, e.g., site is timely handed over, omissions and mistakes in engineering are minimum, free-issued materials, equipment and engineering drawings are timely turned over to contractor, etc.

When the progress of engineering is not sufficiently high and the schedule requirements dictate early start of construction, then selecting unit price basis is inevitable. Contractors, on this basis however, may, tend to perform slower. Moreover, the quantities of work performed require to be carefully measured by spending additional efforts for checking payment reports of the construction contractors.

Hence, the contracting strategy for construction should address the following options:

- Single or multiple contracts
- Scope and price basis for each contract
- Milestone schedules for the entire construction phase as well as for each construction contract
- Extent of material supply by each of the contractors
- Payment terms preferred
- Requirements on guarantees, bank letters, and contractors' insurances
- Liquidated damages for performance and delay
- Temporary facilities
- Construction management
- Third parties to be employed

CONTRACTING PROCEDURE

Following establishment of the construction strategy, the contracting procedure and the list of bidders for each construction inquiry are to be prepared. Contracting procedure is very similar to the procurement procedure described in Chapter 10 – Procurement. It defines how to handle each step of contracting and what techniques and formats are to be used in this phase of the project. More descriptively, it covers mainly the following:

- Main highlights from the contracting strategy
- Pre-qualification requirements and preparation of project contractors (construction bidders) list
- Preparation and issue of inquiry packages including all relevant drawings and specifications, construction requisition, general and special contract conditions as well as bidding instructions
- Pre-bid site visits by bidders
- Obtaining quotations from bidders
- Bid openings
- Technical and commercial evaluations of the bids
- Selecting the successful bidder
- Preparation of letter of intent and formal contract
- Contract revisions
- Invoice checking and payment approvals
- Formats, templates, and documents to be used in contracting

CONTRACTORS LIST

Contractors (bidders) list is prepared in compliance with the construction strategy so that a list of prospective bidders is made ready for each different inquiry planned. The bidders to be included in the contractors list are selected among pre-qualified companies, as in the case of procurement. For bidders, whose qualification dates back more than a couple of years, it is advisable to renew pre-qualification before finalizing the subject list.

Pre-qualification of construction contractors is often made through a pre-qualification questionnaire forwarded by the contracting party and filled in by construction companies addressed. The questionnaire intends to obtain the following information:

- General information on the company including head office organization
- Information on key management staff
- Information on financial status
- Type of works the company has been performing
- Geographical area the company has worked within the last several years
- Organization structure for the contract operations including construction sites
- Quality management system and certificates possessed

- Planning and progress follow-up
- Company's standing on health, safety, and environment
- Engineering capabilities
- Procurement capabilities
- Subcontractors employed
- Construction equipment owned and rented
- Available manpower
- Previous experience and references
- Present commitments

Evaluation of the questionnaire may be followed by a visit to the headquarters and to some of the construction sites of the prospective bidder. If a construction company is pre-qualified through these steps, it can be included in the contractors list. For adding new construction companies to the list, the same steps relevant to pre-qualification will have to be performed, irrespective of the requesting party, whether it is the owner or the EPC/EPCm contractor.

RECEIPT OF BIDS, EVALUATION, AND CONTRACT AWARD

INQUIRY ISSUE

As soon as all the above initial steps are taken, the construction contracting activities can be started by the preparation and issue of the inquiries. It is a good practice to send bidders information about the size and nature of the inquiry and ask for acknowledgment to quote before the official issue of the inquiry. On the one hand, this will help the bidders to get prepared for a soon-to-arrive "Request for Quotation"; on the other hand, it will help the inquirer in estimating in advance the number of bidders that will most probably quote and so, enabling the inquirer to take action, if needed, for securing sufficient number of bids.

RECEIPT AND OPENING OF BIDS

This step is almost identical with what is done in purchasing.

Bidders must submit quotations by the stated bid closing date. Late bids are not accepted unless specifically approved by the project management (and the owner in reimbursable projects).

Bidders are usually required to submit the following:

- Priced original bid required for commercial evaluation
- Unpriced bid copies for technical evaluation

Bidders are often asked to submit quotations in sealed envelopes as the contents are quite critical and sensitive for carrying out a fair and professional contracting campaign. Sealed bid envelopes are kept in a secure cabinet until the opening date. Alternately, bidders may be asked to load their bid on the web to an FTP server by using their passwords assigned for that inquiry.

Bids are normally opened in a closed session one day after the bid closing date by a committee of assigned persons from the engineering contractor. The owner (and/or its PMC may) also participate unless the engineering contractor is executing the project on Lump-Sum Turn-Key (LSTK) basis.

Immediately after the initial recording of the quotation information, unpriced bid copies are forwarded to specialist for technical evaluation. The priced original and copies are kept by the authorized contracting staff for the commercial evaluation. These are handled only by authorized personnel in order to guarantee that the required confidentiality is met.

As stated for purchasing activities, any information or result of bidding or any developments in an inquiry, whether technical or commercial, should never be leaked to anyone, not authorized for the given inquiry. Apart from its importance from ethical point of view, inappropriate handling of quotations in this respect will create risk of receiving no bid responses from the potential contractors during next contracting inquiries.

Bid Evaluation

As in purchasing, the first step in bid evaluation is normally to have the preliminary bid tabulation completed through a systematic review of the bids as received. This enables pre-screening – usually on a price and time of completion level – of the bidders whose quotations are to be further evaluated in detail. Such an overview allows screening out the quotations that are very far from being acceptable due to very high price, very long time of completion, and/or significantly incorrect or incomplete perception of the work.

During detailed analysis, evaluators identify necessary questions and clarifications to be responded by each bidder. Once all necessary clarifications are obtained, the final technical evaluation of the bids can be completed. Simultaneously with the technical evaluation, bids are also commercially reviewed in a similar manner. In case of problems in receiving full satisfactory responses, clarifications meetings may be held with the bidders prior to the issue of the final evaluation. Final negotiation meetings may be held, when necessary, with those bidders that rank at the top in terms of price, schedule commitment and clear interest for the construction work, in addition to full technical acceptability. Such meetings target ensuring the scope of work is precisely understood, all commercial conditions are fully accepted and the final negotiated price is agreed.

When all the necessary inquiry conditioning activities, including final meetings are completed, the Final Bid Tabulation that indicates final prices and the main commercial terms is prepared and issued for approval of the management (and the owner for reimbursable projects). This tab clearly states the recommended bidder to be awarded the contract as being the best bidder of the inquiry.

Contract Award

Construction contracts are awarded to the "technically acceptable lowest priced bidder," which means "the bidder with the total first low cost offering an acceptable time of completion, committing compliance with the negotiated and agreed contract conditions" and most importantly, "deemed to have the capabilities, experience and expertise to perform the works successfully."

Upon approval of the final bid tabulation by management (and owner for reimbursable contracts), a "letter of intent" may be prepared, signed, and transmitted to the selected bidder. It is to be noted that bid tabulations are not presented to owners in LSTK contracts, but only the names of subcontractors are submitted for approval, unless those names are already approved and listed in the contract. Letter of intent normally indicates briefly the total scope and price of the works, time of completion, and the most important clauses relevant to the upcoming official contract. Alternately, the contract may directly be prepared.

Warranty period is mostly set as 2 years from completion of construction. Contractors are required to provide irrevocable and unconditional bank guarantee letters corresponding to some portion of the total order value (e.g., 10%). These bank guarantee letters are kept at hand for any possible liquidated damages for performance and delay and are released when the warranty period expires without any performance issues or delays within that period. Liquidated damages for delay are usually set as 0.5% of the contract value for every week of delay in completion, with a cap of around 5–10%.

Below is a simplified work flow diagram for the inquiry stage showing main contracting/subcontracting activities (Figure 11.1).

Contract (or subcontract) administration mainly consists of the administration and follow-up of the awarded construction contracts at the jobsite. A subcontract administrator is assigned from home office of the EPC or EPCm contractor to the jobsite to deal with the contractual issues that will arise during the performance of the construction works. These issues include claims, extra works or change orders and contractual disputes. In this respect, the role covers contractor's administration of subcontractors which is almost similar to owner's contract administration of EPC contractor, as described in Chapter 18 – Contract Administration. The subcontract administrator deals with the contractual aspects by receiving support from the construction supervisors of his/her team for the technical and performance aspects of the issues. He or she normally reports to the project manager and works in collaboration with the construction manager who is actually the resident head of the construction management team at the jobsite.

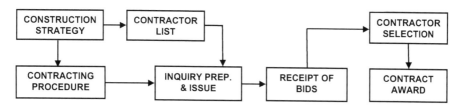

FIGURE 11.1 Simplified work flow diagram for contracting activities during the inquiry stage.

Part IV

Later Stages of Project Development

12 Continuation to the Project at Field

When front end of project services, basically the engineering and procurement have progressed to a sufficient level for construction, it is then time for moving the center of activities to the construction field. Obviously, an industrial facility project is implemented for the core objective of ending up with the desired facility constructed to operate for long years. However, the time and extensive efforts spent until then in making all the necessary preparations for the construction is a crucial prerequisite for a successful construction phase. Such preparatory work basically covers the following:

- Engineering and design for specifying project materials including bulk materials, equipment items, systems and packages to be purchased for installing at the facility
- Specifying the construction through engineering and design for producing the necessary drawings, design details and specifications to be used by constructors
- Carrying out the necessary procurement campaigns to have the project material delivered at jobsite for construction
- Performing construction contracting/subcontracting activities for selecting capable construction companies to perform the construction work

Although it is the construction stage that produces the real tangible outcome of projects, carrying out adequately the prior project stages outlined above as well as in the earlier chapters, is obviously mandatory for achieving the desired outcome. Any significant mistakes, omissions, or deviations made from project requirements during these stages may lead to serious problems including significant delays, cost impacts, and contractual disputes. Such improper execution of a project may also continue causing problems and inefficiencies during the subsequent plant operation and maintenance.

Properly planned and engineered projects enable that the materials and engineering deliverables are timely and adequately supplied and so, the construction teams can better focus on the site work that already involves some other challenges.

Activities associated with construction are normally started much earlier than the construction itself. This is accomplished at the home office during engineering and procurement stages by a group of construction staff who reviews and comments on the design being developed, basically from the perspective of constructability. It is the same group, sometimes called as home office construction discipline, which supports the project team in setting up construction strategy, construction schedule and initial preparations for construction stage, like site surveys.

MOBILIZATION FOR CONSTRUCTION

In construction, the term "mobilization" refers to movement of construction personnel as well as construction equipment (e.g., cranes, loaders, forklifts, diesel generators), construction materials (e.g., scaffolding, ladders, shackles, wire ropes) and tools to the construction site. It also includes, if not already available, construction of temporary facilities consisting of temporary buildings for site management, site fabrication facilities, warehouses, laydown areas, temporary utilities and accommodation facilities for workers. Temporary access roads and site fencing are usually arranged during this period. All the parties (e.g., construction subcontractors, third-party inspectors, construction management team of the engineer) that will get involved in the work at site will have its own mobilization, which actually means getting organized and having the area of works prepared for performing the construction. Detailed plans have to be made for mobilization well before it is started.

Mobilization Plan

Start of mobilization is a major milestone in a project and its timing is identified in the master project schedule in accordance with the status of mainly the following pre-requisite activities:

- Progress of engineering and availability of IFC (issued for construction) drawings
- Delivery status of project materials and equipment

For timing of the mobilization, the project management should ensure that the engineering has sufficiently progressed and the remaining drawings will be timely made ready for feeding continuously the construction work without any pauses that would otherwise lead to costly idle time of the construction workforce.

Similarly, availability of equipment and materials at site in line with the construction schedule should be ensured for avoiding again any costly idle times at the construction field. Delivery time of major equipment and major steel structures requires special attention. The same is valid for delivery of bulk materials, such as piping items. Although it is not required to have all materials at site in the beginning of the construction – and normally, this is not practically possible for fast-track projects – delivery times for bulks, especially the piping material may often be a critical factor in controlling the actual construction progress. The foregoing situation is especially important for fluid processing plants which contain extensive amounts of piping. In order to cope with such concerns, a number of consecutive material take-offs and corresponding orders are placed for piping material items along the progress of piping design.

Timing for the mobilization may also depend on the following:

- Construction strategy (e.g., single or multi-contract construction: The latter allows early start of construction at site with the initial portions of the work, such as site grading or civil works that require relatively limited progress in engineering and procurement.)

- Type of project (e.g., grass roots or brown field facility: The former may necessitate early start due to some additional initial works.)

When a multi-contract scheme is decided to be applied, it is often started with a separate site grading or civil works contract. For such an approach, it is sufficient to have only those drawings basically for soil works (excavation, landfill, foundation locations, etc.) available as IFC. This type of construction contracts sometimes includes in its scope certain other related works, such as installation of grounding grid and sometimes, construction of temporary facilities as well as permanent buildings; hence, engineering on these items should be available for such construction work. These would still form a limited portion of the entire drawings of the process facility. Accordingly, it is possible to go out for construction bid requests and contract awards much earlier in multi-contracting approach.

As stated above, the start of mobilization may also depend on the type of the project. If it is a new (grass roots) plant, early start of construction for making at least part of the temporary facilities ready for certain initial works (e.g., drilling water wells, perimeter fencing) and new surveys (e.g., additional geotechnical investigation, mapping) may be decided. For remote and unsecure sites, facilities for security forces may first to be constructed before the other temporary construction facilities. These are generally performed by small local subcontractors that can quickly mobilize. On the other hand, for a revamp (brown field) project in an existing facility, owner's plans for turnaround (shutdown of live units) imposes an important timing criterion for the project. Mobilization time is set up accordingly, noting that mobilization should be started sufficiently in advance of the scheduled shutdown date in order to perform the necessary prior work that should be executed before the shutdown, like prefabrications for piping, structures, and equipment foundations.

MOBILIZATION OF CONSTRUCTION PERSONNEL

Construction personnel to mobilize to a jobsite basically belong to one of the following parties:

- Owner's construction (follow-up) team
- Project Management Consultant (PMC) or Owner's engineer
- EPCm or EPC contractor and its construction subcontractors
- Construction contractors (when owner selects and directly manages such contractors)
- Third parties (e.g., for inspection, Health, Safety, and Environment audits, training)
- Vendor supervisors (of complicated equipment/systems during installation, pre-commissioning, and commissioning)

Owner's construction follow-up team should always be present in the picture during construction, especially for approvals. A PMC may or may not have been employed by the owner depending on owner's strategy for management of the works. A contract on EPCm basis also means owner has given the contractor certain responsibilities for acting on behalf of itself. All these parties perform follow-up of construction contractors and subcontractors and in that sense, they represent construction management staff of the project. The situation of an EPC contractor is not much different, as it requires performing good management services at site for its construction crews as well as its subcontractors.

At the start of construction, the first assignees from the owner (or its representatives) are usually a small team under management of a construction manager supported by few team members. The team is responsible for making the arrangements required for easy and fast mobilization of the construction contractors and also performs monitoring, controlling, and decision-making functions during such initial periods. As the construction works progress, more team members from necessary disciplines may be added to the team, as needed.

A similar approach is followed by each construction contractor and subcontractor. Their team, composed of few staff and direct laborers in the beginning of mobilization, is expanded to a full team of construction workforce, with the number of employees ranging from few hundred to several thousand at the peak, depending on the size and schedule of the project.

MOBILIZATION OF CONSTRUCTION EQUIPMENT AND MATERIALS

For performing the construction work, contractors require moving to site various construction equipment and materials in accordance with the scope and schedule of the works. The construction equipment items include the following:

- Excavators, loaders, trucks, graders, cranes, forklifts, hoists, post-weld heat treatment equipment, welding machines, radiographic filming devices, hydrotesting pumps, drain pumps, diesel generators, temporary electrical cabinets, etc.

Construction tools, appurtenance and consumables to be supplied by contractors during mobilization are usually the following:

- Scaffolding, hand tools, gaskets, timber for temporary supports, formwork, lifting accessories like wire ropes, slings, shackles; consumables like welding roads, grinding disks, radiographic films, grease and oils, cleaning material, safety gadgets, etc.

TEMPORARY FACILITIES

Temporary facilities at a construction site consist of temporary buildings, fabrication shelters, workshops, warehouses, laydown areas, mess halls, dormitories, temporary roads, and utilities. These are almost always arranged by each contractor for its own use, often with provisions for the owner's site personnel.

Temporary buildings are for use as offices, meeting rooms, dining hall, dormitories, toilets, and bath rooms for both the owner's (including owner's representatives) and contractor's personnel at site. The buildings are furnished with the equipment and furniture required for the intended service. For project materials, temporary laydown areas and warehouses are needed. For large material items, like large-size valves, fittings, flanges and pipes, laydown areas are to be prepared with sort of simple fencing around. Warehouse is used for relatively small-sized materials such as small fittings, gaskets, bolts and nuts as well as for sensitive and fragile items like instruments. Temporary shelters may be needed for prefabrication of mainly the piping spools and steel structures.

Water and electrical power connections to the temporary facilities are made by the contractor who will use these utilities during the entire construction stage. Tidying up and cleaning all the temporary facilities as well as sterilizing and keeping especially the dining halls, bath rooms, and the toilets to match proper hygiene requirements for human health are handled by the contractor at its own care and cost.

Sources of water and electrical power are usually provided to contractors by the owner. However, contractors are often asked to provide their standby equipment for power and install their distribution systems needed for performing their work.

CONSTRUCTION

Construction is obviously the stage when all efforts and money spent until then for obtaining the required deliverables through engineering and procurement, starts generating a more tangible product, the industrial facility desired. At this time, the epicenter of project activities is relocated to the field, though there still remains some home office support required to the project in engineering and procurement. The project manager of the EPC or EPCm contractor is still the head of his/her project organization and may move to and settle at site depending on the nature and criticality of the project. Alternatively, he or she may, instead of settling at site, make frequent visits to site for monitoring and control of the progress and attending project review meetings. Top management continues to keep an eye on the project, mostly through its nominated project sponsor, especially with regard to contractual standing of the project, including expenditures versus budget, progress versus schedule and potential risks, such as major disputes and liquidated damages.

Some more highlights for a typical construction stage of work is given in Chapter 13 – Progress of Project at Field.

CONSTRUCTION MANAGEMENT

Building an industrial facility is an expensive investment. Such a decision is taken only if detailed feasibility studies and subsequent reviews reveal an attractive level of profitability. This obviously depends also on the conformance of the construction with the project requirements. Accordingly, the owners do not rely only upon how well prepared the construction contracts are, but tend to have sufficient monitoring and control either by themselves or through their PMC for avoiding any risks to the project. Similar concerns lead EPC or EPCm contractors to pay close attention to the progress of construction works by way of close monitoring and control of the construction crews and the subcontractors at site. All these interests necessitate setting up effective site management organizations by these parties to drive the construction stage of the project as proficiently as possible. The functions and contribution of the construction management to the project is further outlined in Chapter 13 – Progress of Project at Field.

DEMOBILIZATION OF CONSTRUCTORS

Demobilization is moving away the manpower and removing the temporary facilities that are not required to be on site anymore at the time when the construction approaches to its end. However, part of the labor, temporary facilities and construction equipment are still to be kept at site until end of commissioning due to possible modification works that may be necessary until the plant is fully in operation. Such work is mostly minor modifications in mechanical, electrical, and instrumentation-control works that may arise for correcting or fine tuning the operation of the facility. Contractor's manpower in other disciplines may also be kept at site if the contract allows to have minor non-critical (i.e., no effect on operation and HSE) works not fully completed at mechanical acceptance, like some portions of paving, insulation, and painting.

13 Progress of Project at Field

Construction phase of a project is literally started when the project's first construction contractor begins mobilizing to the jobsite. However, the construction work actually commences at the time when the construction contractor sufficiently mobilizes to site with labor, materials, construction equipment and tools and begins with the execution. At this time, the construction management team of the owner (and if exists in the project, its Project Management Consultant/Contractor (PMC) or Engineering-Procurement-Construction management (EPCm) contractor who is in charge of following up the construction work on behalf of the owner) actively participates by monitoring and managing the construction works by the construction contractor.

CONSTRUCTION ACTIVITIES AND ORGANIZATION

Construction is part of the backend works of a project either performed by an EPC or EPCm contractor selected by owner (or its PMC) or directly contracted out by the owner to a construction company when the owner is acting alone. The scope of construction works may be split into more than one contract depending on the construction strategy set up during the earlier phases of the project as mentioned in Chapter 11 – Construction Contracting. In terms of different trades of construction works, the entire job can be divided into the following disciplines:

- Site preparation/site grading works
- Civil, structural, and building works
- Structural steel works (sometimes included within mechanical works)
- Mechanical works including equipment erection
- Piping works (sometimes considered to be part of mechanical works)
- Instrumentation and control works (sometimes considered together with electrical works and named as E&I works)
- Electrical works
- Other works that are either included in one of the above contracts or separately contracted out, like insulation, refractory lining, fire proofing, painting, etc.

SITE PREPARATION/SITE GRADING WORKS

These works refer to all earthworks including excavation, backfilling, leveling, compaction, and testing. The scope of these works also includes activities like measuring, surveying and setting out, planning and drafting. The execution is carried out in accordance with the site preparation drawings that are mainly based on the

159

topographical survey data, geotechnical report, and plant layout. Site grading is not often limited to scraping only the vegetated soil layer on the surface and adjusting to the grade levels specified, but also comprises digging and removing some meters depth of soil to be replaced and compacted with stronger land fill material. During these works, it may be more practical to make also some major underground installations, like main sewer system and general plant grounding system.

CIVIL, STRUCTURAL, AND BUILDING WORKS

These works normally consist of performing a variety of concrete works, including structure foundations, equipment foundations and saddles, slabs, plinths, concrete structures, pipe sleepers, retaining walls, trenches, underground galleries, buildings, etc., all with relevant tests and usually with construction contractor-supplied bulk materials. Reinforced concrete works require preparation of re-bars and formwork before pouring concrete for foundations, structures, sleepers, retaining walls, galleries, and trenches. Curing time of concrete is an important factor in proceeding with the subsequent works. For most common concrete mixes, it takes weeks to fully cure unless a curing aid is added to reduce it down dramatically. After sufficient curing of the concrete, the formwork is taken out. Paving and final coats of roads are usually constructed at later stages in order to avoid possible damages during movement of construction equipment. Until then, a temporary base preparation followed by gravel laying is usually sufficient. Regarding the buildings of the plant, if the building is multi-floor and the structure is concrete, then a step-by-step construction is to be followed starting with foundations, then the columns of the grade floor and then beams and slab of the upper floor, repeated for subsequent floors. Building works also require masonry, plastering and other fine works as well as mechanical works such as plumbing, HVAC, and firefighting; electrical works such as power distribution, grounding, data network, lighting, fire detection and alarm.

STRUCTURAL STEEL WORKS

These works refer to fabrication and installation of structural steel for canopies, steel buildings, skids, pipe racks, supports, loading platforms, stairs, ladders, etc., often prefabricated and delivered to site with primer paint or hot dip galvanized coating. Fabrication of structural steel is mostly performed at contractor's workshop. Materials and consumables are usually under contractor's scope of supply. The most common steel structures in process plants are the pipe racks located at the off sites, process units and the interconnection of these plant sections. Steel structures are also used when modularization is applied to process or utility systems by arranging the equipment items with interconnecting piping, cables and instruments on skids for shortening erection time. Buildings of an industrial facility may be selected to be in steel. Despite having notable advantages of using steel as the structure (e.g., short erection time, robustness, flexibility in making modifications), it has the disadvantage of being weak during fires. This requires application of fire proofing in order to reduce such weakness, at least to delay possible collapse before which some precautions, such as immediate cooling, evacuation of personnel, and extinguishing, can be taken.

MECHANICAL WORKS

Mechanical works primarily include installation of equipment, some of which are entirely designed and supplied by specialized vendors, while some may be fabricated by the mechanical contractor. The former items are pumps, compressors, turbines, and other specialty vendor designed items as well as pressure vessels and heat exchangers, most of which require specialist manufacturers. Mechanical contractor-fabricated items are mostly site-fabricated tanks and workshop-fabricated simple vessels. Equipment erection involves lifting of heavy and large items followed by installation and alignment. Assembly of equipment internals is also performed by mechanical contractors. The works usually cover piping and ducting systems and their connections to equipment. These systems require to be prefabricated before being erected, post-weld treated (if necessary according to the codes), supported and tested at the jobsite. Equipment items and piping materials are purchased by EPC/EPCm contractors or owners. Painting and insulation of the piping and ducting as well as equipment are often included in mechanical contractor's scope if not separately contracted out. In addition to the works in process units, mechanical contractors also construct utilities (water, compressed air, steam, cooling media, fuel, inert gas, etc.), storage tanks and fire protection system (consisting of fire water pumps, foam system, piping network with hydrants, monitors, deluge and sprinkler systems, tanks). HVAC and sanitary systems of buildings are also part of mechanical construction work.

PIPING WORKS

Piping works, often performed under the category of mechanical works, can as well be carried out as a separate contract. For performing the piping works of a facility, contractor will need to prefabricate the pipe spools in contractor's external workshop or alternately at site workshop built as part of the temporary facilities. Piping material consisting of pipes, valves, fittings, flanges, etc., have to be ordered much before start of the construction contract in order to have these materials timely delivered to the site or to the workshop for prefabrication. Prefabrication of piping involves cutting pipe to required dimensions, beveling, welding fittings, flanges, and other pipe to form pipe spools which are to be fabricated to transportable dimensions to the erection area. Prefabrication at the workshop provides more comfortable and better controlled working conditions that enhance the quality and the speed of production, compared to piping erection directly at the construction point. The spools are then brought to the location of installation, usually as primer coated, and then fitted there by pipe-fitters. This is followed by welding of all weld passes by pipe welders. Small sized piping (i.e., less than two inches) are usually erected directly at the erection location without prefabrication. Such an approach may be selected as erection of small size piping takes much less time and also allows adjustments in the place against minor deviations that might have been present in the drawings or created during erection of other items.

Prior to any welding activity, Welding Procedure Specifications (WPSs) and Procedure Qualification Records (PQRs) are to be prepared by the contractor and submitted for approval to owner appointed quality control/inspection staff. All pipe welders are to be qualified prior to being allowed to perform any welds. Qualification testing of welders is usually done at the jobsite in the presence of owner appointed inspection personnel. Welds are subject to non-destructive examination (often radiography) according to the criteria, procedures, and percentages set-forth by applicable codes. Piping loops are then pressure (hydraulically) tested by the contractor after assembled and fully welded.

INSTRUMENTATION AND CONTROL WORKS

These works consist of installation of instruments (e.g., indicators, transmitters, switches), junction boxes, Programmable Logic Controller (PLC), Distributed Control System (DCS), relevant panels and cabinets, instrument cables, tubing with compression fittings (for instrument air), cable trays, conduits, etc. Instrumentation works also include loop checks, calibration of instruments, tests and inspections. Installation of certain in-line instruments, like control valves and motor-operated valves, as well as safety relief valves and level gauges, is usually handled by mechanical contractor of the project. Electrical and piping bulk material of instrumentation and controls, such as cables, conduits, tubing, compression fittings, supports and the consumables for the installation is usually supplied by instrument erection contractor, while the tagged items, like instruments, cabinets, etc. are purchased by EPC/EPCm contractors or owners. Part of these works are at the field (referred to as field instrumentation), mostly within the process and utility units. These are related to installation of field instruments with wiring to junction boxes that are then cabled to instrument panels. The remaining works are mainly related to power supply to instruments and control system, automation and controls (PLC, DCS, and Safety Instrumented Systems) for which the panels and relevant software loaded computer workstations are commonly located in an instrument control room. Instrument cables are routed through conduits and cable trays or ladders. For easy identification purposes along the life of the facility, contractors are asked to label cables, junction boxes, cable terminations, panels and instruments. As some process plants treat flammable or explosive substances, electrical instrumentation and the electrical installation for such facilities as well as the corresponding materials should comply with the hazardous zone classification made during the engineering.

ELECTRICAL WORKS

Electrical construction works include primarily the installation of electrical power equipment, like transformers, switchgear, motor control centers, uninterrupted power supply; installation of electrical systems, such as power distribution, grounding, lighting, lightning protection, communication and cabling for all these systems. As most electrical equipment items are long delivery items, these are ordered by EPC/EPCm contractors (or owners) well before electrical construction works start. The rest of the materials, mainly the bulks, like cables, conduits, cable trays, ladders and glands as well as lighting fixtures are usually left to electrical contractor's scope of supply.

Certain areas of the project site may have been classified as "hazardous" against the risk of fire and/or explosion. The classified zones are shown on "hazardous area classification drawings" prepared during engineering in line with applicable codes. Electrical materials like junction boxes, lighting fixtures, safety switches, local control stations, emergency stop pushbuttons and their cable glands to be installed in classified areas should be suitable to the zone classification in accordance with applicable codes.

Constructor's Site Organization

Constructor's site organization is normally headed by a resident site manager (named also as construction manager) and supported often by a deputy site manager. Site management staff of the constructor supervises its direct and indirect personnel. The following direct construction manpower constitutes the persons who work straightly on the construction activities:

- Field superintendents (when there are different work areas)
- Foremen (the number depends on the disciplines of work within the scope as well as the number of teams formed for each trade of work)
- Skilled personnel (welders, pipe fitters, bar benders, riggers, construction equipment operators, etc.)
- Semiskilled personnel
- Helpers

The site personnel who are essentially at the backstage to direct, monitor, control, and serve the direct manpower are normally considered as indirect construction labor. These persons are mostly site office staff, like construction manager, site engineers and draftsmen, site buyers, accountants, coordinators, supervisors, planners, time and cost keepers, etc.

It is the amount of direct manpower of a contractor that receives a very major attention in verifying the actual progress of the works with the plan. These progress values are continuously followed up and periodically reported for each trade of the works in order to enable taking necessary measures before facing with any notable delays against the planned progress figures. Parameters other than levels of direct manpower, such as the efficiency of performance, may also be a problem in an unsatisfactory performance of a contractor. This may arise from inadequate management and coordination (indirect manpower), unsuitable work conditions and lack of sufficiently competent labor. Hence, all these factors have to be continuously monitored and controlled for keeping in track the progress of construction.

Practices Supporting the Construction Operations

Apparently, availability and good quality of construction drawings, specifications, construction material, equipment, and labor are not sufficient alone to guarantee a successful construction stage. Successful construction means ending up with a fully completed facility in compliance with the design, the program, and the budget without any health, safety, and environment (HSE), contractual and legal issues. This obviously requires also good management, coordination, and supervision which are only possible if certain professional practices are set up and applied all along the entire project execution period. These practices are mainly sort of reference tools, plans, and procedures for monitoring and controlling the construction operations and require to be strictly followed at field. Such supporting practices are often enforced by the contracts. Some examples are as follows:

- Programs, plans, and procedures
  - Construction schedules
  - Manpower histograms
  - Construction organization charts
  - Execution plans
  - Site coordination procedure
  - Change management procedure
  - Interface management plans
  - Work permit, "PTW" procedures
  - Method statements
  - Pre-commissioning plans and procedures
  - Mechanical completion plans and procedures
  - Material control procedures
  - Material preservation procedures
  - Heavy Lift plans
  - Site logistics plans
  - Simultaneous (concurrent) operation plans
  - Commissioning plans and procedures
  - HSE plans
  - Quality assurance and control plans
  - Site inspection and testing procedures
- Reports, studies, and logs
  - Monthly, biweekly, and weekly progress reports
  - Constructability studies
  - Lifting studies
  - Periodical manpower and safety reports
  - Incident and accident records
  - As-built marked-up drawings
  - Construction dossiers
  - Test and inspection reports
  - Certification requirements for material, workmanship, and tests
  - Daily logs of direct and indirect labors at site
  - Daily logs of construction equipment and tools at site
  - Logs of unworkable days
- Notes and notices
  - Minutes of construction kickoff meeting
  - Minutes of weekly and daily site coordination meetings
  - Minutes of schedule and progress meetings
  - Minutes of safety meetings

- Minutes and registers of interface management meetings
- Non-conformance reports and registers
- Notice to proceed for hold points
- Extra work approval requests
- Claims
- Notice for mechanical completion
- Various acceptance certification forms

CONSTRUCTION MANAGEMENT ACTIVITIES AND ORGANIZATION

In principle, it is primarily the owner who essentially needs to put in place a construction management organization. However, the perspective of owner for management of construction is inherently different from the perspective of EPC and EPCm contractors since owners do not normally provide direct supervision for the works. These services are contractually assigned to EPC/EPCm contractors and the owner has to act rather as an auditor as well as the approval authority in the project. For following up the works more closely, owner may delegate such duties and authorities to a PMC as to represent the owner in front of the contractors. This approach is selected especially when the project is large and/or complicated.

As cited in previous sections, the principal leader of a project holding the entire responsibilities in that project is the project manager (or a project director). The leader manages all the phases of the project with support from his/her staff to whom functional duties, responsibilities, and authorities are delegated within the project organization. Both the owner (and PMC, if exists in the project) and contractors have their own project managers assigned to represent their sides during the execution. This is how engineering, procurement and construction contracting activities are carried out. Similarly, construction management follows the same rule and the top management position of the project is still held by the project manager, but he or she delegates construction management functions to his/her construction manager during that stage. Such a management configuration may as well be established by owners. The site chief position of each of these parties is typically named as construction manager. Meanwhile, EPC and EPCm contractors have to assign a full construction management organization with engineers, coordinators, and supervisors for each discipline of work present in the scope of the construction.

The duty of field construction management organization of a contractor is to ensure a professional construction stage by supervising and coordinating all field operations in accordance with the contract as well as with good engineering and construction practices commonly accepted in the industry. The organization basically focuses on assuring the works are executed in full compliance with the project documentation, the contract and the applicable codes, standards, and regulations, by giving always the priority to avoiding any possible HSE incidences. Achieving a successful construction stage during which the works are completed within budget and on time with a good HSE performance is the main target for all the field staff involved. A typical organization of a project management team during construction stage is shown in Figure 13.1 and of a construction management team is indicated in Figure 13.2.

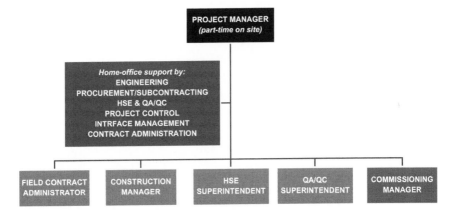

FIGURE 13.1 A typical project management team organization during construction of a process facility.

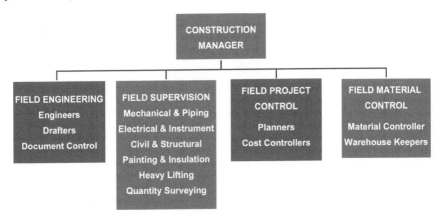

FIGURE 13.2 A typical field organization of construction management team of a process facility project.

FINAL ACTIVITIES AT FIELD UP TO ACCEPTANCE

As construction approaches to its end, finalizing the arrangements for the last stages of project activities becomes critical. The objective is to timely facilitate the plant handover to the owner and to obtain the official acceptances. These final efforts are outlined in the following paragraphs.

PRE-COMMISSIONING AND MECHANICAL COMPLETION

"Mechanical completion" is a key milestone of a project at the end of construction and is contractually documented with a certificate signed by the owner. It declares completeness and robustness of the plant (or section of it for which the mechanical completion certificate is drawn) in line with the project drawings, specifications, applicable codes, and the contract.

The definition for the extent of mechanical completion may change from one contract to the other, based on the fact whether it includes (and overlaps with) pre-commissioning or not. Project owners often tend to include pre-commissioning in mechanical completion, though contractors prefer to exclude for having earlier achievement of a possible payment milestone with mechanical completion before pre-commissioning is totally done. Accordingly, the next step after mechanical completion becomes either pre-commissioning or directly commissioning. For the former case, mechanical completion is reached by construction-type employees with participation of vendor supervisors, as required. However, for pre-commissioning and commissioning, experts specialized in that field have to take over, while only some key construction personnel are still kept on standby for any possible dismantling and re-erection requirements. The most common approach for projects in significant size and/or complexity (e.g., oil and gas projects) is to have pre-commissioning done for the mechanical acceptance.

Mechanical completion including the pre-commissioning refers to construction, fabrication, assembly, and non-functional testing to confirm integrity of the entire construction of a plant or a section of it. For mechanical, structural, and piping portions of the project, it includes visual inspection for complete and correct installation. Vendor representatives participate for large and/or complicated equipment and system packages during installation and pre-commissioning. For tanks and vessels, internal inspection is to be made as well as hydrostatic testing. For piping, non-destructive examination and hydrostatic testing has to be completed with pipe supports and hangers in place. Other tasks may include steam blowing, air blowing, flushing, chemical cleaning, drying, and bolt tensioning. All items subject to damage during flushing, cleaning, and pressure testing are to be removed during these activities and to be reinstalled afterwards. For electrical systems, verification of proper installation of cable trays, supports, cables and wiring, cable bends as well as continuity, insulation, terminations, and grounding is to be made and validated. Electrical parts of instrumentation and controls undergo similar type of inspection with the electrical systems. Additionally, instruments and instrument valves require to be checked and calibrated. Instrument tubing has to be cleaned, flushed and pressure tested. In practice, some minor works (like painting) that are not related to HSE and plant performance can be transferred to the next phase without being fully completed. At this milestone, mechanical completion certificate is accompanied by a punch list indicating such incomplete minor items. Mechanical completion with a punch list is probably a good way to balance the interests of the owner and contractor and it is a commonly used practice.

Pre-commissioning part of the activities involves non-operating adjustments and cold alignment checks without introducing process and/or auxiliary fluids and without permanently energizing the plant. Some of the activities mentioned above, such as flushing, blowing, and chemical cleaning are considered to belong to pre-commissioning. The plant is often divided into easily manageable and independent sections or system packages for pre-commissioning and commissioning. In fact, such a split should normally be decided and documented well before mechanical completion so that it can be achieved system by system, to give way from mechanical completion to commissioning.

Commissioning and Start-Up

Commissioning is the phase in which the process fluids (and solids such as catalyst or raw materials, if any) are charged into the plant. During this phase, electrical equipment items and components are energized. For sections of the plant running flammable substances, commissioning may need to be started by using inert fluids first. Equipment and piping are purged and pressurized. The plant or sections of it undergoes operational trials during which running adjustments and functional checks are made. Commissioning is a major phase transition of a project which represents handing over the facility by the construction team to the commissioning team.

An important prerequisite is to have all commissioning procedures prepared and owner's approval received before transition from construction to the commissioning. During the same period, training of the owner's engineers and operators are to be completed as well. In order to start commissioning of the process units, all utilities have to be commissioned and started-up to be operationally ready for introducing the actual fluids in the process.

At the successful completion of commissioning, the operational efforts can be switched to the start-up of the plant with process fluids already loaded (in oil processing plants, this achievement is often referred to as "oil-in") and process conditions adjusted for making the products. This is the starting point for generating revenues after long periods of time elapsed from the initiation of the project along which the owner had to continuously spend money as capital expenditures.

Finally, before permanent commercial operation is actually started, the plant is tested for guaranteed performance dictated by the contract.

ACCEPTANCE PROCESS

At a certain moment in time after completion, the plant has to be taken over by the owner. Usually this event coincides with also the acceptance of the plant by the owner. It is not possible to describe the acceptance process in a single straightforward manner, as it depends a lot on the type of contract, and specifically on the way the contract defines the takeover. In the most general cases, the plant takeover may be at the occurrence of one of the following events:

- Mechanical completion: It is when the plant has been completed according to the contractual specifications and the pre-commissioning has been successfully executed. At the time the contractor deems that the condition of mechanical completion has been reached, it issues a request for mechanical completion certificate to the owner. If the owner is also satisfied that the contractual conditions are met, then it will issue the certificate, discharging the contractor's relevant responsibilities under the contract. If, on the contrary, the owner believes that such conditions are not met, it may object and not issue the certificate, instructing contractor to remedy the owner identified defects that prevent the plant from being considered as completed. If, on the other hand, the conditions preventing the plant from being considered as 100% complete are minor, consisting of unsubstantial deficiencies, then the owner may issue the certificate with a list of such minor works still to be completed by the contractor. This list is usually known as a punch list, or butt list. One of the conditions for this approach is that the remaining defects will not inhibit safe execution of the further phases.

- Ready for commissioning: This condition does not differ very much from mechanical completion, in that the plant should have reached mechanical completion and should be ready for the introduction of process fluids (and solids, if any).
- Provisional acceptance: This means that the plant is not only mechanically completed but has also been successfully tested for performance and found in line with the contractual requirements. If the plant taking over is at provisional acceptance, then the mechanical completion punch list must have already been completed and the relevant minor defects rectified.

When the plant is accepted as described above, takeover is reached and the owner takes full responsibility for care, custody, and control of the plant and starts operating it with its own resources. In some cases, contractor may be requested to provide operation and maintenance services as well, but this is generally done under a separate contract.

The last of contractor's liabilities are discharged during a period of time following provisional acceptance, during which the plant is operated normally and the contractor is still responsible for failure of any plant components. Contractor has the obligation to repair or replace parts that break down or go wrong due to previously unidentified initial defects. The duration of this period – usually known as mechanical guarantee period – is generally in the range of 12–24 months and requires to be negotiated during contract award, as the owner will normally want a longer term, while contractor prefers a shorter one. Such an approach by contractors is not only due to an obvious desire to reduce its responsibilities, but also because of the fact that the corresponding back-to-back guarantees obtained from vendors of the equipment and materials have limited validities starting with purchase order dates. Accordingly, contractors do not like to have their guarantees extended beyond such periods.

For a detailed description of the plant acceptance, discharge of guarantees, and relevant liabilities from a contractual point of view, please refer to Appendix 4 – Contractual Clauses.

Part V

Project Completion

14 Contract Closure

A contract is like a living creature with a certified birth date (the effective date), certified events (milestones, deliveries, payments, changes, etc.), and an inevitable conclusion. Like in the case of human beings, the conclusion also has its course of events and its certifications.

Now, let us consider what the aspects to be formally closed in a contract are as follows:

TECHNO-ADMINISTRATIVE CLOSE

PERFORMANCE OF OBLIGATIONS

It is essential to certify whatever is included in contractor's scope of work was actually executed and completed and is free from defects, to the extent specified in the contract. This is done by means of certificates issued at different steps, such as mechanical completion, provisional acceptance, and final acceptance. The latter is the final certificate, attesting that contractor has performed to meet all requirements that the contract specifies as its responsibility.

This series of certifications attests not only that things have been done, but also that they have done properly. What exactly does "properly" mean? The answer is: There is no general, absolute definition of the "good" as opposed to the "bad" performance; there is one and only one meter of judgment and this is the contract. Whatever the contract states that "shall be done," should actually be done, not more, not less. In the old philosophical diatribe between "perfect" and "perfectible," perfect represents the ultimate result of an improvement process that outcrops achieving the best ever or possible (from Latin perficere, to accomplish, to complete, to execute), whilst perfectible represents something that tends toward perfection but it is at an intermediate stage. Human activities are perfectible, everything we do can be improved, but trying to obtain absolute perfection is a waste of time and energies. Therefore, when a contract defines targets and guaranteed specifications, meeting them is to be considered as contractually "perfect," yet things are still perfectible because there is always a way to improve them. Based on this perspective, nobody could demand a contractor to obtain better results than those specified in the contract, even though they can theoretically be improved.

PAYMENTS

A contractor should testify that all the moneys due to the contractor under the contract have been duly received. If not, there must be a reason like deductions and similar, which must have a contractual justification. Missing this, the contract cannot be considered closed.

BACK-CHARGES

Back-charges fall under the category of payments; they are payments from contractor to owner for matters like damages, penalties, reimbursement of costs borne by owner out of contractor's fault and the like. Back-charges must have been either paid by contractor to owner, or deducted from payments due by owner to contractor.

BANK GUARANTEES

Bank guarantees must be null and void and their original copy must have been returned by owner to contractor. One could argue that once the guarantee is expired and its value has been zeroed, it cannot be enforced against the guarantor (the contractor). Yet financial directors become very nervous when the originals of the bonds are not returned, because in some extreme cases they can become a liability and so they are considered as a loose cannon and a potential threat.

CLAIMS

As discussed in Chapter 18 – Contract Administration, a claim is an intended entitlement that a party deems to have vis-à-vis the other. Once a claim is filed, there is always the possibility that money will change hand and so the economics of the contract cannot be closed. Therefore, in order to attest that the contract can be closed, it is necessary that all claims are settled in a way or another: accepted and paid, withdrawn, awarded by a court or an arbitration tribunal and paid.

FINAL CLOSING CERTIFICATE

All the above items must be included in a final certificate, that some call a techno-administrative close out certificate, whereby both parties certify that they are satisfied and there is no other pending issue, nor obligation outstanding from the other party. This is also called a "waiver" by some project owners. The parties should also expressly declare that they have nothing else to demand or request of the other. This is sometimes called a tombstone agreement.

OTHER CLOSING ACTIVITIES

The techno-administrative close out certificate has the purpose to put an end to the contract management between the parties. In order for either party to consider the contract completed, usually there are other requirements from the company corporate system. In many companies, a contract cannot be declared as closed unless and until these requirements have been satisfied.

LESSONS LEARNED

It is a document prepared by the project manager with the cooperation of the project team, analyzing the performance of the contract from both sides, with the aim of leaving a testimony for future use. The lessons learned are of two types.

Errors Made

It is about things that could have been managed better but for some reason went wrong, causing damages or embarrassment to the company. This analysis must be deep, thorough and above all, sincere. The point is not putting the blame on anybody, but rather analyzing the reasons that led to a negative performance and identifying what the remedial actions could have been. The past is past and this report is not the right place to complain or punish anybody, but the managers of future projects can highly benefit from a good lesson learned report. Often, a good execution plan to be prepared since the proposal stage for a project in the same country, or with the same client, or with other similar characteristics, can start from the lessons learned report of a previous one. It should not be forgotten that next project can happen after several years, or the people who have followed the current one may leave the company, so that the only practical way to hand down experience to posterity is a good lesson learned report.

Better-than-Planned Performance

In many cases, during the performance of a contract, new ways, techniques, or procedures are identified or experienced and they prove beneficial. The reason could be ingenuity of the project team, suggestions by owner, cultural mix with a joint venture partner and so on. Whatever the reason is, these areas are also of fundamental importance, as they allow improvement of the company's best practices and must be made available to the company so that it can take advantage of them in future projects.

PROCUREMENT AND CONTRACTING FEEDBACK

Many companies (both project owners and contractors) have procedures and databases in their procurement and contracting system to provide feedback about suppliers and contractors. Owners require feedback about contractor and likewise contractors desire to have their own on their suppliers and subcontractors. The main purpose is that next time a procurement or a contract/subcontract manager will have to draw up a vendor or subcontractor list for another project, he/she will automatically find the feedback from previous ones, which will be a valuable tool for decision-making, both in the positive and in the negative sense.

A particular case of this database is the collection of information that could make a given supplier or contractor not eligible for future projects; usually this refers to black lists for misbehavior (bad faith commitments, lack of responsiveness, lack of ethical behavior, etc.). Generally, a vendor or contractor who has been found responsible for these kinds of misbehavior can be black listed for a certain period of time, or – less frequently so – indefinitely. Reinstatement back into the vendor list is usually subject to some bureaucracy (reports, certifications, etc.) to make sure that, to the best of one's knowledge, problems have been mended. Black listing of vendors or contractors is often regarded as a very private and confidential information for a company and is not pronounced officially in any environment other than to related staff within the company.

OTHER PROJECT CLOSING ACTIVITIES

For an industrial facility project, a contract is a control tool that dictates owner's requirements including liabilities of the contractor against the owner. However, certain actions or activities of the contractor have nothing to do with the contract between the owner and the contractor. These are based on the contractor's operational requirements, practices, standards, procedures, policies, work instructions, guides, etc. Examples regarding the project close out, not directly related with the contract closure, are as follows:

- Dismissing the project team, task forces, and terminating the employment of the agency personnel and free lancers
- Stopping or limiting man-hour charges to the project by the disciplines
- Putting all records and communications in an auditable form
- Filing all final engineering, purchasing, quality and construction documentation including the as-built that may later be needed
- Sale, if possible, of surplus material
- Preparing project close out report
- Generating the final cost report of the contractor's company that shows actual expenses including make good and contingency amounts

Part VI

Other Functions with Top Priority

15 Health, Safety, and Environment

The outlook on health, safety, and environment has continuously been evolving and attracting more considerate attention ever since the development of the industrial society that basically started with the industrial revolution. Expectations of the society for a safe and healthy living without damaging the environment while enhancing the quality of life through industrialization have led to due care on these conceptions. The requirements have progressively been developed and imposed by governments and international organizations through laws, codes, standards, and regulations as well as international protocols.

For Health, Safety, and Environment, the acronym "HSE" is very commonly used, but there are also others in use, like the following:

- EHS
- SHE
- HSSE (the second S stands for "security")

For many of the industries, the care for and improvements in HSE are regarded as one general function that are made up of these three interrelated subfunctions. This is the case for production industries in project implementation, plant operation, and maintenance. However, especially in terms of legislation, these are usually regulated by two or three different governmental administrations based on the particular limits of duties assigned to such administrations. Furthermore, the ever-increasing recognition regarding the need for environmental protection (that had previously been significantly neglected) leads to focusing environmental issues separately in many state and private organizations.

BRIEF HISTORY OF EARLY HSE LEGISLATION

Booming complaints for miserable working conditions and accidents in the production industries brought occupational health and safety issues to the forefront of the society. The first laws were a series of UK labor laws, known as Factory Acts, passed by the parliament in the early years of the 19th century. The early acts focused on working hours of the children in cotton mills, but were not effectively enforced until the Act of 1833. The regulations on working hours were then extended to cover women about a decade later. Meanwhile, public rage that arose from the dangerous working conditions and frequent accidents in coal mines resulted in the Mines Act of 1842. Early environment laws were also issued in the UK around 1850s as a response to the great stink that arose due to dumping sewerage into the river Thames. Another landmark development was the release of Clean Air Act due to the Great Smog suffered in London in 1950s.

Few decades after mid-19th century, legislative developments in HSE started to emerge also in United States and continental Europe. Establishment of ASME as a standardizing body was one of the pioneering events in United States that was initiated as a response to high fatalities in the industry. The first legislation on social insurance and workers' compensation in Germany around 1880s is another example to early advancements triggered by public unrest on working conditions.

KEY HSE TERMS AND DEFINITIONS

Some of the important terms frequently used in HSE terminology are briefly defined below.

HAZARD

The potential to cause harm, including ill health and injury; damage to property, plant, or products or the environment; production losses or increased liabilities or loss of reputation.

HAZARD IDENTIFICATION

Determining which hazards exist or are anticipated with their characteristics and possible outcomes.

RISK

Possibility of suffering harm or loss; the likelihood that a specified undesired event will occur due to the realization of a hazard by or during work activities or by the products and services created by work activities. It is a combination of the frequency (probability and exposure) and the consequences of the occurrence of a hazard.

RISK ASSESSMENT

A systematic process of organizing information to support a risk decision to be made within a risk management process. It consists of the identification of hazards, its analysis and evaluation associated with exposure to those hazards.

INCIDENT

An instance of something happening; an event or occurrence.

ACCIDENT

An undesired and unexpected event mainly caused by human error (and/or high-risk conditions that results in or has the potential for physical harm to persons and/or damage to property and/or interruption of business).

Top Event

The "release" of a hazard leading to the undesired event at the end of the fault tree or at the beginning of an event tree or at the center point in a bow-tie diagram (aforementioned techniques are described in next paragraphs of this chapter).

Near-Miss

An unplanned event that threatens HSE or assets wherein any protective barrier that may be present is challenged, but is somehow avoided, so that it remains as just an incident that does not turn out to be an accident.

Barrier

A precaution taken for preventing the release of a hazard or for providing protection once a hazard is released. These are either tangible, like shields, isolation, separation, protective devices, or non-tangible, like procedures, warning signs, and training.

Control Measure

Barriers put in place to prevent the threat from releasing the hazard, or to restrict the consequences of the top event from escalation.

IDENTIFICATION OF HAZARDS AND ASSESSMENT OF RISKS

There are numerous techniques used in various business sectors for qualitative and quantitative investigation of hazards and risks. Both the qualitative and quantitative types aim finding a link between a hazard that would affect a system and a failure of its individual components. Both of these types question possible causes leading to release of a hazard; however, the quantitative approach also goes for estimating probabilities and severity of consequences.

The techniques used for risk assessment primarily target to precisely identify all cases of hazards through realistic scenarios, including those with effects on environment. Some of the basic traditional methods for safety analysis used broadly in various applications are listed below:

- What-if
- Checklist
- FMEA (Failure Modes and Effects Analysis)
- FTA (Fault-tree Analysis)
- ETA (Event-tree Analysis)

What-If

This technique requires going through the flow sheets and operating procedures for considering what would happen if unexpected situations happen. The answers can lead to discovery of potential problems. It is a qualitative method whereby potential consequence of deviations is analyzed.

CHECKLIST

It is made up of a list of items to be checked on areas of known concerns. It cannot identify issues that have never been met, so often used together with what-if technique to enable analyzing from a wider perspective.

FAILURE MODE AND EFFECTS ANALYSIS

FMEA is a bottom-up approach that intends to identify critical component failures which may cause catastrophic consequences. Therefore, the analysis begins with identifying ways that each component can fail. Then effects of these failure modes are reviewed.

FAULT-TREE ANALYSIS

FTA is a "top-down" method for analyzing system failures. It begins with identifying first a possible catastrophic failure or top event that should be avoided. Then component failures that could lead to or contribute realization of that top event are analyzed. It can either be qualitative or quantitative.

EVENT-TREE ANALYSIS

ETA is a "bottom-up" technique that starts from an undesired initiating cause (system or component failure) and proceeds with possible further system events that lead to a series of final consequences. The course of the event is not analyzed. The method can be used both quantitatively and qualitatively. It can be combined with quantitative risk analysis (QRA) and fault tree analysis.

HSE IN PROCESS FACILITY PROJECTS

During execution of projects, minimizing risks and avoiding hazards are the prime objectives above any other interests. These objectives are to be strictly pursued all along engineering, manufacturing, construction, commissioning, and other specialty services including the subsequent plant operation and maintenance. For all these services, plans, work procedures, and methodologies are to be devised as strictly HSE compliant. Accordingly, the core responsibilities in HSE are vested in line management of the above outlined functions and disciplines taking part in those activities. Meanwhile, HSE staff involved has the vital role of acting as consultants and auditors in the identification, assessment, and follow-up of the hazards and the precautions. In other words, building and operating a safe, healthy, and environmental friendly plant is the duty of the engineers, designers, constructors, and the operators, while HSE staff participates by way of monitoring, assessing, and commenting on any HSE issues defined during assessments and analysis.

HAZARD MANAGEMENT

Hazards are to be effectively managed to eliminate or at least minimize personnel exposure, injuries and loss of life, environmental pollution, loss of business, and loss of reputation arising from accidents that can reasonably be anticipated. This can be achieved by reducing the risks that may release a hazard leading to an accident to as low as reasonably practicable (ALARP) level starting from the engineering and design stage. It is accomplished by setting out the following objectives:

- Eliminating hazards where possible
- Minimizing the probability of the release of hazards
- Preventing escalation (reducing effects) of a released hazard
- Protecting persons from exposure to hazards

Engineering and design is the most effective phase for implanting HSE compliance to process plants while developing the projects. This is achieved through having each engineer and designer strictly follow, during design, the applicable references that mainly consist of the following:

- Laws (e.g., labor laws, environmental laws)
- Codes (e.g., pressure equipment directive—PED, machinery directive—MD, ATEX, building codes)
- Local regulations (e.g., on emissions, fire protection)
- International standards (e.g., ANSI, ASME, ISO, EN, NFPA)
- Specific HSE standards (OHSAS 18001, ISO 45001, ISO 14001)
- Company standards (i.e., engineering company's or owner's standards, procedures, guides, etc.)
- Contract scope and requirements associated with HSE
- Good engineering practices (i.e., methods and applications commonly accepted and widely used in the industry)

All these references are used to prepare an elaborate plan for HSE specific to the project. Within the last several decades, it has become customary to prepare an HSE plan or hazard management plan at the initiation of each project. Such a plan covers all phases and stages of the project from its initiation, up to the handover to the owner. The plan is supplemented by relevant procedures for the activities and operations that have the potential of releasing hazards along any stage of the entire project execution. HSE plan outlines mainly the implementation of HSE processes for providing the following:

- HSE compliant development of engineering for the all stages of the project, with a sharp focus on subsequent plant operation
- Safe and healthy working conditions for all persons involved with the execution of the project and operation of the plant
- To meet environmental goals with minimum possible environmental footprints
- Security to all persons involved, material delivered and the assets built

Inherently Safer Design

Inherently Safer Design (ISD) represents a specific outlook to design for achieving the HSE goals in process plant projects. It targets achieving the following:

- Much less hazards by way of eliminating those that are practically possible to remove from the project (e.g., isolating hazardous areas, minimizing the sources of hazard release)
- Minimizing the number of people exposed to hazards by reducing as much as possible the manual activities (e.g., increasing the level of remote control and automation in hazardous areas)
- Reducing the inventory of hazardous material and so the severity of any top event (e.g., minimizing stored volumes of hazardous substances)
- Simplicity and reliability to reduce risks
- Use of passive protection to maximum possible extent (e.g., using equipment and systems with fully rated pressures instead of installing relief devices)
- Minimum dependence on active protection systems
- Minimum dependence on procedural safety systems

HAZARD IDENTIFICATION AND ASSESSMENT

Systematic studies for identifying, assessing, and analyzing hazards and risks are carried out during all stages of a project. Project HSE reviews or hazard management activities normally start at the initiation stages of project and proceeds along its development through Front-End Engineering and Design, detailed engineering, construction, pre-commissioning, commissioning, and startup. It is often extended to early stages of plant operation and maintenance as part of the project, so-called as the post-commissioning. For process industries, the most widely used techniques are as follows:

- Hazard Identification (HAZID)
- Bow-Tie
- Hazard and Operability Analysis (HAZOP)
- Quantitative Risk Analysis (QRA)
- Layer of Protection (LOPA) and Safety Integrity Level (SIL) Assessments

HAZID is one of the most commonly used techniques having appreciable flexibility and ease of use. It is often applied at the initial stages of risk management activities. It is a combination of identification, analysis, and brainstorming based on the hazards listed in predetermined checklists. It serves to identify hazards and potential mechanisms that may lead release of the hazards. It helps designating the necessary barriers while portraying the main risks for a project or a facility by providing recommendations to reduce or manage those risks. The facility is divided into separate sections, and the team starts identifying and evaluating potential hazards for each section. Then possible causes that may lead to hazards, likely effects (consequences), and associated safeguards/controls already in place are spotted. It can indicate further needs for more elaborate studies. A similar technique used for identifying environmental impact hazards is named as Environmental Impact Identification (ENVID).

Bow-Tie method is based on a diagram that shows the threats and consequences on just one sketch as shown in Figure 15.1. The diagram is shaped like a bow-tie, with a number of potential threats on the left-hand side of a node at the center, tied to those bundles of multiple threats. The threats are the hazards that can release the top event resembled by the central node. On the right hand side of the node, there is the bundle of possible consequences, all tied up back to the central node, completing the bow-tie appearance of the sketch. Additionally, there are barriers in the bow-tie located on both sides of the top event directly on the ties to the node. Barriers at the left of the node function to control the threats for avoiding the top event to occur. Barriers at the right of the node work to eliminate escalation into the undesired consequences. Major accident hazards can be analyzed using the bow-tie methodology after completion of HAZID. It is used to provide assurance that the hazards can be managed to levels that are ALARP.

HAZOP is the most widely used hazard analysis technique in process industries. It is "outside the box" type brainstorming method for identifying and resolving hazards by considering also apparently unusual occurrences. Although it is a bottom-up technique, it is more efficient than FMEA as it can dismiss earlier the component failures that have no effects to system operation. First step is identifying the normal operating condition. Next, a guideword is used to go over a possible deviation in the process. Different guide words (such as "no," "less of," "more than," "reverse," and "other than") outlined in Table 15.1 are used to identify potential deviations in a system with this qualitative method. HAZOP review sessions are primarily based on the P&IDs.

QRA is also a commonly used method in the process industries. In order to conduct a QRA, detailed information derived from previous analysis, like HAZOP, FMEA, FTA, ETA, etc., is necessary. QRA is not a single method, but a continuous advancing of previous analysis results on a more detailed level. An important characteristic of QRA is the F/N curve produced, where F stands for the frequency of the damaging event and N stands for the number of victims. It is a formal systematic approach to estimating the likelihood and consequences of hazardous events by expressing the results quantitatively as risk to people, to environment,

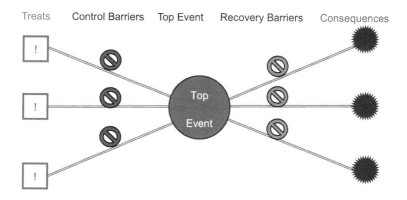

FIGURE 15.1 Bow-tie method.

TABLE 15.1
HAZOP Guide Words and Deviations

Guide Word	Description of Deviation	Possible Deviation
No/None/Not	The intended/specified condition is not achieved at all	No flow, no specific component present, etc.
More of/ Higher than	Increase in quantity/level of a design parameter	More flow, higher pressure, higher temperature, higher concentration, etc.
Less of / Lower than	Decrease in quantity / level of design parameter	Less flow, lower pressure, lower temperature, lower concentration, etc.
As well as/ More than	Some additional/unspecified/ unintended condition happens	Unexpected components, extra phase, etc.
Part of	Only part of specified condition occurs	Change in ratio of or missing components, partial flow, etc.
Reverse	Reverse of what is intended or specified happens	Reverse flow, reverse change in temperature or pressure, etc.
Other than	Any other deviation from normal operation	Completely unintended substances, phases, operation, etc.

to company assets and reputation. QRA is widely used in assessing the risk in oil and gas installations, especially in refineries, tank farms, cross country pipelines, terminals, etc.

LOPA analysis and SIL classification aim evaluating and strengthening the plant's safety that is based on safety instrumented functions (SIF) and a safety instrumented system (SIS, formerly named as emergency shutdown, ESD). LOPA is a semi-quantitative technique for analyzing process hazards. It is started from the results of HAZOP to treat each identified hazard with regard to initiating cause and the protection layers (e.g., basic control system, alarms, operator intervention, SIS, relief devices, physical protections) that prevent release of a hazard. It assesses the layers of protection in place, defines the SIL and shows if more risk reduction is required.

The above are only some brief highlights about the techniques used in hazard management. Further details about these techniques are a matter of special expertise and are outside the scope of this book.

HSE AT SITE DURING CONSTRUCTION AND COMMISSIONING

The project HSE plan may include all the requirements applicable to site work or alternately a separate site HSE plan may be prepared to supplement the former plan. Such site HSE plan requirements are basically as follows:

- HSE responsibilities at site and site HSE organization
- Managing of subcontractors from HSE point of view
- HSE monitoring and reporting requirements
- Simultaneous (concurrent) operations

- Communication
- Incident reporting, investigation, and corrective actions
- Inductions, training, and competency
- Inspections, audits, and corrective actions
- Change management
- Health, hygiene, and medical facilities at site
- Medical examinations for fitness to work
- Emergency response
- Safe working (HSE) procedures

The safe working procedures to be prepared and pursued during site work normally cover the following:

- Permit to Work (PTW) system
- Lifting operations and heavy lifts
- Working at height and dropped objects
- Scaffolding and ladders
- Personal protective equipment (PPE)
- Construction equipment and operators
- Driving and transportation
- Safety signs
- Hand and power tools
- Excavations
- Pressure testing
- Welding, cutting, and grinding
- Noise and hearing conservation
- Radiation sources
- Chemicals and hazardous materials
- Compressed gas cylinders
- Fire prevention and firefighting
- Electrical safety and energy isolation
- Emergency response
- Site security
- Housekeeping and waste management

Personal Protective Equipment

The minimum requirements for PPE are mainly the following:

- Non-metallic hardhat
- Safety boots with toe and soles protection
- Goggles for eye protection
- Ear plugs for hearing protection
- Gloves
- High visibility vest
- Full length pants and long sleeve shirt
- Miscellaneous other protections required for particular works

All personnel at site requires to be trained for performing safely and environmental consciously the tasks they are responsible for. Training services are provided by the contractor to its own personnel and subcontractors' personnel for all job disciplines.

All personnel of the contractor and subcontractors have to attend an induction (safety orientation) prior to starting the work at site. The induction normally covers the following topics:

- Overview of construction activities and hazards
- Applicable HSE plans and policies, legal requirements
- Permit to work system
- Hot work
- Fire protection
- Working at heights
- Scaffolds
- Confined spaces
- Excavations
- Electrical safety
- PPE
- Construction equipment safety
- Site traffic rules
- Emergency response including evacuation and mustering
- First aid
- Housekeeping
- Site safety meetings and toolbox talks
- Disciplinary procedures and safety incentives

Toolbox talks (daily work pre-start briefing) are held by the attendance of contractor's and subcontractors' personnel as well as their immediate supervisors or foremen before starting each shift. Such informal meetings intend to have the workers focus on the job ahead in terms of safe work practices and are arranged to take less than a quarter of an hour. Apart from toolbox talks, the contractors and subcontractors are responsible for attending scheduled HSE meetings.

During site activities, HSE compliance is monitored by HSE site superintendents who support construction and commissioning on HSE matters by making daily inspections with the authority to stop any HSE incompliant work. In fact, any person who notices an HSE risk during any work being performed at site has the responsibility of warning to have it stopped. All contractors and subcontractors at site are required to report all HSE incidences and observations at site.

Each contractor, project owner, and plant have HSE recording and reporting system used to follow up and control the HSE status of the project and generate HSE statistics. During evaluation of construction proposals, these statistics from previous jobs constitute an important criterion for HSE performance of contractors and subcontractors. HSE criteria mostly reported are as follows:

- Incidents and accidents
- Near-misses

- Recordable injuries
- Workdays lost due to occupational health and injury
- First-aid cases
- Property damages
- Fire incidents
- Fatalities

Safety has long been regarded to have the top priority in all types of work and has been described with the motto, "Safety First." Occupational health and environmental protection have been accepted to bear similar crucial importance a little later. In short, no concerns about any contractual liabilities, loss of time and monies can justify HSE incompliance.

16 Quality Management

"Quality management" has noticeably developed within the past century due to increasing needs of the society for ensuring that products and services consistently meet customer requirements. The foregoing situation applies similarly to the services, documentation, materials, fabrications, and constructions produced during industrial plant projects.

QUALITY MANAGEMENT SYSTEM

In order to comply with the expectations of the society for quality, producers of services and products devise for themselves quality management systems (QMSs) that fit to their objectives and operations. A QMS is focused not only on quality of products and services, but also on the means to achieve it since the former cannot be guaranteed irrespective of the latter. Hence, it is a philosophy and strategy designed to involve everyone who participates in the production and consumption of the products or services, namely

- All members of the organization having the responsibility in the process of producing quality products and services, including the management and the workforce
- Suppliers
- Contractors
- Customers

It is important to note that quality management (QM) can only be powered by top management's declaration of the commitment to it as a corporate policy.

QM can be regarded as an outcome of the culture, attitude, and organizational structure of a company that aims providing products and services to the satisfaction of its clients' needs. For industrial plant projects, the clients are those people or entities referred to as owners or investors throughout this book.

The culture requires quality in all aspects of the operations, with work to be done right at the first time, by eliminating as best as possible any defects and wasting of resources (Figure 16.1).

Accordingly, QM makes strong focus on data collection, evaluation, and the control of the process. It has four main components:

- Quality planning
- Quality assurance (QA)
- Quality control (QC)
- Quality improvement

> *Work is to be done right*
> *the first time*

FIGURE 16.1   One of the basic principles of QA.

QMS, therefore, uses QA and QC of processes as well as the consequential discoveries to achieve more consistent quality. Although such systems require to be tailor-made for each company to fit each particular organization, all systems generally have in common the following elements:

- The organization's quality policy and quality objectives
- Quality manual
- Procedures, instructions, and records

The QMS standards created by International Organization for Standardization, ISO, are for certifying the processes and the system of an organization, not the product or service itself. The principles focused are basically as follows:

- Meeting clients' requirements and endeavoring to exceed the expectations
- Establishing, by participation of leaders at all levels, common purposes, targets, and harmony to have all get engaged
- Enhancing the organization's capability to produce and supply valuable products and services by competent and committed individuals at all levels throughout the organization
- Ensuring consistent and foreseeable results are obtained efficiently by way of fully understanding and effectively managing processes and arranging these as a coherent system
- Continuously focusing on identifying and applying improvements
- Performing systematically and persistently data analysis and evaluations for basing the decisions on producing desired results
- Maintaining and managing necessary connections with other stakeholders, such as subcontractors, vendors, and customers for sustained achievements

QM is an approach that organizations use to improve their internal processes. A successful application environment requires a committed and well-trained workforce that effectively participates in quality improvement activities. When it is properly implemented, this approach can lead to decreased costs, better overall performance and quality products or services, and an increased number of happy and loyal clients.

ROOTS OF QM

QM is literally a recent phenomenon. However, its roots can be traced back centuries along the development of societies and civilizations in the sense that clients of craftsmen had been keen to look for quality. This led craftsmen to train and supervise their apprentices and get organized in guilds. Regarding the roots of QM, some have a different view and state that it had begun even much earlier when humans started taking notes after the discovery of writing.

It was only after the Industrial Revolution that the early QMSs were initiated as sort of standards that controlled process outcomes. As more people had to work together to produce more and more, best practices were needed to ensure quality results. Eventually, best practices were established and documented forming the basis of standard practices for QMSs.

On the path, arriving at today's QMSs, there have been a number of management gurus taking stage. Some of the most outstanding pioneers and the advances they brought to life are as follows:

- Frederick Winslow Taylor (1856–1915), a mechanical engineer who sought improving industrial efficiency. He is sometimes called "the father of scientific management." He was one of the intellectual leaders of the efficiency movement and part of his approach laid a further foundation for QM, including aspects like standardization and adopting improved practices.
- W. Edwards Deming (1900–1993), Joseph Juran (1904–2008), and Armand V. Feigenbaum (1922–2014) were noticed by Japan as quality gurus and were asked for help to fix poor quality of Japanese consumer goods. It was after end of World War II when Japan was directed to change its focus from being a military power to an economic one. Japanese were trained by and followed the instructions of these management experts who worked independently from each other. The focus widened from quality of products to quality of all issues within an organization. The outcome was a big success. In the 1950s, QC and QM developed quickly and became a main theme of Japanese management. In the 1970s, Japanese products were begun to be seen as leaders in quality.

In the 1980s and 1990s, a new phase of QM, known as Total Quality Management (TQM), began as the American contribution to the new quality movement started in Japan. Having observed Japan's success on employing QM principles, Western companies commenced introducing their own quality initiatives. TQM, developed for the broad spectrum of quality-focused strategies, programs, and techniques that embraced the entire organization, became the center of focus for the modern quality movement. It was around those times that the term, "Total Quality Management" lost its fame and the term "Quality Management System" or "QMS" has become its replacement for representing the latest form of the quality management systems.

SYSTEM STANDARDS AND TECHNIQUES

There are a number of system standards and technics used in setting up and improvement of quality. ISO 9001 is by far the most recognized and implemented quality management system standard in the world. It specifies the requirements for a QMS that organizations can use to develop their own programs. Other ISO standards related to quality management systems include the following:

- The rest of the ISO 9000 series (including ISO 9000: Basic concepts and Language and ISO 9004: Guidelines for performance improvement)
- ISO 10005 (quality management systems-guidelines for quality plans)
- ISO 14000 series (environmental management systems)
- ISO 19011 (guidance on internal and external audits of quality management systems)
- Various other ISO standards for particular production industries, like medical devices, automotive-related products, food industries, etc.

ISO 9001 is one of a series of standards that can be used for external QA purposes. It specifies quality system requirements for use where a contract between two parties requires the demonstration of a supplier's capability. In other words, it serves in circumstances when conformance to specified requirements is to be assured by the supplier during several phases of activities which may include design, development, production, installation, and servicing. It should be noted that these requirements are complementary to the technical specifications of the product. They do not replace the technical requirements and are not alternatives to them.

When an organization's quality system has been assessed with reference to ISO 9001 by an accredited independent certification body, then the quality system is registered, and can be used as evidence of QA in tendering for contracts. Quality systems produced in accordance with these quality system requirements are subject to regular third-party assessment based on documented and objective evidence of compliance.

Dealing with accredited suppliers provides their clients a sense of security, and reduces the effort required to control the supplier's products. From the supplier's point of view, accreditation generates a quality image, customer confidence, and access to markets where quality certification is obligatory. In addition, the introduction of a QMS will have a major positive effect on internal performance of entities that apply it.

QMS IN PROCESS PLANT PROJECTS

QMS in industrial plant projects targets building a plant completed in a manner that results in safe, reliable, maintainable, economic, and problem-free operation. It also focuses on minimizing rework, delay, and or downtime during execution of the scope of the project.

Prominent contractors in the process plant business have already been operating on a sort of quality system since long time ago, though it was not then named as TQM or QMS and no quality certification had then been available. The particular methods and tools used were their own standards, procedures, and practices generated upon need and based on accumulated experience. This type of business was in fact obliging contractors to base all project decisions and engineering selections on strong criteria recorded and documented with corresponding owner approvals due to the following:

- Complexity of these multidiscipline specialty jobs that require effective exchange and incorporation of correct information with quite a number of parties involved all along the project execution.
- Crucial need for establishing and maintaining relatively long-term relations with the owner, subcontractors, and vendors in effective coordination and collaboration all along the project.
- Possible future disputes with owners or other stakeholders on project decisions and selections made.
- Probable claims that may be raised by owners or local authorities due to alleged nonconformities as well as possible claims by other project participants for possible changes that may have developed along execution of the project.

Within the past decades, it has been almost an obligatory practice to have a well-structured and certified QMS applied to the services of each function and discipline of the parties involved in process facility projects. The parties are mainly the engineering contractors (basic engineering, detailed engineering, FEED, EPC, or EPCm contractors), PMC, owner, vendors, construction contractors, and subcontractors. Each has its own system for performing the services and QMS documentation that consists of a manual and company's standards, procedures, work instructions, guides, forms, and such other related reference tools. All parties carry out their internal operations in accordance with their own systems; however, for any matters that can lead to incompliance with owner's standards with regard to some details, formats as well as material and construction features, owners may request in the beginning of projects to have their own standards followed. Normally, the contractors are ready and flexible enough to accept and adapt themselves to making such modifications in their operations according to different clients when requested.

Regarding application of QMS in process plant projects, the reader should keep in mind the general steps mentioned in the beginning of this chapter and indicated in Figure 16.2.

A brief verbal description could be that QA covers the methodologies applied for managing the quality of execution. On the other hand, QC refers to the techniques and procedures to be used for verifying the quality of the output. These two, together with the quality planning and quality improvement, are referred to as QM.

DUTIES AND RESPONSIBILITIES

For an engineering contractor performing engineering, procurement, and construction (e.g., on EPC or EPCm basis), the responsibilities of the functions taking part in the project are outlined in their QMS documentation. In other words, all the functions participating directly in the project execution for engineering-design, material-equipment supply, planning-cost control and subcontracting as well as project and construction management perform their duties in accordance with the

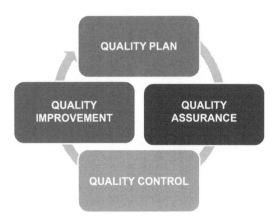

FIGURE 16.2   Basic components of QM.

corresponding job descriptions and work instructions included in the company's quality manual. It has become customary that the contractors prepare project-specific QA/QC plans and procedures for each project they execute. The responsibilities, reporting lines, and the QA/QC activities planned for the project are stated in this document and the project team members are presented on organization charts proposed for the project. The QA plans describe the processes for assuring the scope of work and the entire project implementation comply with the contract, codes, standards, and good industry practices. QC involves verifications of the work done by all the project functions by way of checking for approval or rejection and includes also inspection and test plans (ITPs) to be used during fabrication and construction. QM is not an exclusive responsibility of the QM staff. It is the direct functions which bear the responsibility for carrying out the Q planning, Q assurance, Q control, and Q improvement. They are from the project operation/execution groups, such as project management, contract management, engineering, project control, procurement, construction subcontracting, construction management, construction, and commissioning.

On the other hand, the company's QM staff has particular roles and responsibilities in the project and act by way of monitoring and reporting to have the required assurance and control of the quality in place during the project execution. They also have the duty of compiling the information obtained from their internal audits as well as from the feedbacks received from the project functions. Hence, it can be predicted that the responsibilities for quality in project execution basically lies with the line management of the direct functions while the quality engineers and managers act as consultants and auditors. In other words, the QM team should always have unbiased freedom and authority in the implementation of the QM plan, free from schedule, cost, and any other concerns. QM staff assigned to a project may normally report the PM (and CM at site), but should mainly have primarily a direct and strong reporting line, without any intermediary, to the corporate assigned management level responsible for the QM. In this respect, these QM specialists act similar to the

HSE specialist in a project. In fact, it should be noted that QM has strong links with HSE in projects.

METHODS AND TOOLS USED FOR QM

The entire team assigned to the project is responsible for the quality of the project. The same is also valid for the indirect company functions, such as general services, human resources, accounting and top management in bearing their portion of responsibilities under the company's overall QMS. Their roles are mostly to provide support to project execution, often not by direct participation in the QA/QC activities of a particular project.

The basic methods and tools used by the direct project functions are as follows:

Project Management

Project Manager (PM), as the head of the project team, bears the overall responsibility for application of the QMS in the project. Accordingly, he or she has the duty of planning, organizing, monitoring, and controlling of all the project activities with particular attention to QA and QC. Additionally, he/she has to prepare or have his/her specialists in the team prepare for the QA the following project documents:

- Project Coordination Procedure
- Project Execution Plan (PEP)
- HSE Plan
- QA Plan (if not entirely included in the PEP)
- Criticality rating assessments for project equipment, materials, packages, and subcontracts
- Change management and control procedures
- Correspondence with owner and other parties on contractual matters, including the relevant registers
- Minutes of meetings with owner and with others on the project
- Progress Reports

The above is just a brief outline for the major points that have ties with quality. The entire duties and responsibilities of a PM are described in Chapter 6 – Project Management, in detail.

Engineering

Each engineering discipline described in Chapter 8 – Engineering follows certain methods and techniques specified in the QMS for assuring and controlling the quality of their services in the project. The engineering procedures are integral to the contractor's QM plan. The activities of each engineering discipline in this respect are basically as follows:

- Ensuring compliance with the contract requirements, applicable codes, standards and regulations as well as other QMS requirements.

- Preparing the project specifications to describe the requirements regarding equipment and material supply as well as construction.
- Setting up and incorporating in the requisitions for equipment and packages the vendor data and drawing requirements.
- Carrying out checks on calculations, drawings, data sheets, specifications, and all other engineering-design outputs generated by the discipline; signing each after being reviewed for endorsing as checked and approved before release.
- Interdisciplinary circulation of design documents and information for comments; incorporation of agreed comments in related engineering deliverables.
- Participating formal design reviews (P&ID, plot plan, MOC, model, HSE, SIL, etc.)
- Attending project meetings as well as site visits as required.
- Checking technically vendor bids and vendor drawings for compliance with the project requirements.
- Verifying the engineering and design also through manual calculations, comparing with previous similar designs and alternate calculation methods.

The above is a brief outline for the major activities of engineering disciplines in relation to QA and QC of the services. The entire duties and responsibilities are described in Chapter 8 – Engineering.

Procurement and Construction Subcontracting

Procurement and construction subcontracting activities are described in Chapter 10 – Procurement and Chapter 11 – Construction Contracting, respectively. A brief listing of the activities related to QMS can be outlined as follows:

- Preparation of bidders list in line with the work instructions included in the QMS.
- Carrying out prequalifications for prospective bidders with which there has not been either any or any recent direct experience.
- Preparing and issuing the inquiries in accordance with the procurement and subcontracting procedures.
- Collecting bids, carrying out bid evaluations, placing orders and performing order follow-ups until end of and acceptance of delivery in line with the applicable procedures for all materials and equipment. Doing the same in construction subcontracting in line with the project procedures.
- Arranging inspections and expediting for vendor supplies; issuing procurement reports (e.g., material progress summary, inquiry list, order list, inspection/expediting reports) in line with the relevant procedures.
- Checking the quality certifications of vendors, subcontractors as well as the certificates for material/equipment with respect to the project requirements.

Construction Management, Construction, and Commissioning

For these activities, details are given in Chapter 12 – Continuation of the Project at Field, and Chapter 13 – Progress of Project at Field.

Regarding such site work, the following particular methods and tools are used in accordance with the QMS:

- Constructability reviews made during home office services represents one of the important design verification activities for the quality as well as the viability and safety of construction.
- Work permit application procedure (often called as Permit to Work, PTW) and the information to be submitted for the permit request, like method statements, represent enforcements for good work planning, and verification to bring in quality as well as safety of the work to be done.
- NDT (non-destructive testing) activities are performed in accordance with the contract requirements, codes, engineering standards, and good engineering/construction practices which are outlined in and often referred to as inspection and test plan, ITP. Results of these inspections and tests (and the subsequent corrective actions for failed tests) stand out as an evidence for the quality of the construction.
- Mechanical completion is an important milestone in the project as marking the time of transition from construction to commissioning. It represents the entire plant (or the declared portion of it) is mechanically complete in accordance with the approved procedure and is ready for commissioning and startup, in conformity with the quality and safety requirements of the contract.

Nonconformities

QMS requires raising Non-Conformity Reports (NCR) for any situations identified along execution of the project services as conflicting with the system requirements. This applies to mistakes, omissions, deviations, and delays that would affect the quality of project implementation. NCRs are raised by any of the members of the project team, not only by the quality engineers who follow up the project performance from the quality point of view. NCRs are registered and followed up for prompt treatment and resolution. These are reviewed in periodical QA/QC meetings together with other quality subjects such as results of quality audits.

17 Job Acquisition Process

Scope of this chapter is to show the complete cycle that aims at and leads to the acquisition of new jobs, the lymph of every company that operates by projects, as opposed to routine sales of products.

The cycle is basically composed of the sales and the proposal processes. This chapter does not intend to provide a thorough description of either process, but rather aims showing the entirety of the cycle and the interactions between the two groups. Therefore, this book will neither enter into the meanders of the sales job, nor into the technicalities of the proposals. Both sides require specific preparation and the professional profiles of the managers and employees on both sides are slightly different; the two systems need to interact very tightly for the success of company's initiatives.

KNOWLEDGE IS POWER

Engineering and construction contractors' business is characterized by a somewhat short period perspective. If the business of a project owner is considered, say in refining or petrochemical, it can easily be assumed that the whole business (from the first idea, down to the decommissioning and dismantling of the commercial installation) spans over a period of time of easily 30 to 40 years. Therefore, everything in the project owner's management should be set to match this timing, from market strategies to product strategies, from financial strategy to manpower planning and so on.

Out of this period, what concerns the Engineering and Construction (E&C) contractor is usually the first few years, i.e., from the first idea to the completion and final acceptance of the plant. One could argue that if just proposal and plant implementation are analyzed, this period of actual operating contractor's involvement would be even shorter. Nevertheless, limiting our attention to such a reduced span would be strategically wrong and quite risky. Many contractors are essentially market reactive, limiting themselves to responding to market solicitations like Requests for Proposal (RFP) and concentrating on how to prepare the best proposal for that specific piece of work. Unfortunately, such a strategy can work only if the market is in a bullishly enthusiastic phase and the contractor has a dominating position for aspects like technology, territorial presence, or exclusive arrangements with a big important customer. In cases like these, a contractor can probably sit and wait for RFPs and its main concern will be to submit a competitive proposal and win the job. It will hardly need a sales and marketing structure to anticipate the RFPs.

In a more general case, experience shows that a contractor has an interest in having prior knowledge about the projects, therefore, involving itself in the very initial stages during which the project owner conceives the project. This would give several advantages, like the following:

- Create a database of accessible projects, even though they have different degrees of probability to become real and a different level of interest for the contractor.
- Make more accurate financial planning, predicting volumes of work and even profits with a large time in advance – essential for a company quoted in the stock market.
- Improve manpower planning, not only in quantity but also by skill, in order to capture the indications of the market about the types of work to come and the relevant technological fields. For example, knowing that a certain type of technologies will tend to be utilized in the next future, would help an engineering contractor selecting the most adequate profile for the recruiting of its process engineers.
- Even in a fat cow type of market, such an anticipated outlook would allow contractors to select the most interesting projects and even afford them the luxury to become choosy, concentrating efforts on those projects which appear to be more in line with one's experience and position, for matching strategies better.

What is, therefore, needed to put such a system in place?

THE SALES FUNNEL

A funnel is the best representation of the way a contractor sees the market opportunities:

- A far end that is very wide and includes many projects, as many as it can be visualized for the coming future in a span of say, up to 5 years. In the far end, there are projects generally still ill-defined and with various probabilities or degrees of interest, but you do have to be aware of them, because one never knows, a not too interesting project could evolve into an interesting one, for many reasons.
- A close end, much narrower, which includes few selected projects that are about to materialize and have higher probabilities of success for the contractor. At the close end, there are mostly projects that are already at the bidding stage, for which the contractor considers itself to have good chances of success.
- In the middle, there is a process of narrowing of the funnel, by discarding some projects, either because they get cancelled or awarded to others, or because there are too little chances to pursue them successfully.
- In general, by moving along the axis of the funnel, the projects contained at various levels will become fewer in number and more probable.

Let us categorize projects according to their probability and let us utilize the following categories:

- High probabilities, those on which the plans are based. They are the "must win", in order to meet the expectations of the shareholders

- Medium high probabilities are those that are not considered as "must be won" at that moment, but the interest must remain focused on them, because, in the event that a high-probability project is lost (or canceled), then it becomes crucial to work intensely on the others in order to replace the lost ones
- Medium probabilities
- Low probabilities

The question that would come to everybody's mind is now: How can a project's probability be improved? The main answer is "by working on it". Sales people generally visit potential customers trying to promote their company, convince them of the advantage they would bring and obtaining information on the prospective project. This information is used by the contractor to take all those actions that would improve chances of a good execution. For example, allocate best resources in advance, recruit the right people if necessary, make alliances and pre-bid agreements, fine tune the contracting strategy, and so on. These actions are time consuming and have a cost; therefore, the effort is concentrated on – although not limited to – the most probable projects.

The above contains the answer to "how to improve the probabilities for getting awarded a project". If one of the high-probability jobs is lost, then it should be decided which of the medium high ones should replace it and then sales efforts are to be concentrated on it, in order to create the conditions to make it more probable and "upgrade" it to "high".

Figure 17.1 will explain how a sales funnel works, by giving a graphical representation of the sales funnel of a hypothetical contractor in a given moment.

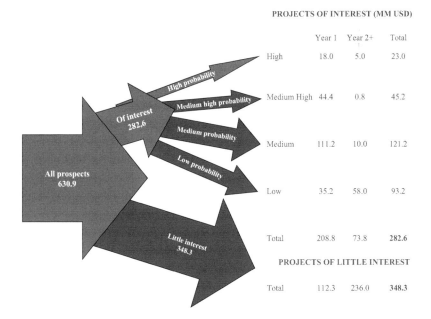

PROJECTS OF INTEREST (MM USD)

	Year 1	Year 2+	Total
High	18.0	5.0	23.0
Medium High	44.4	0.8	45.2
Medium	111.2	10.0	121.2
Low	35.2	58.0	93.2
Total	208.8	73.8	**282.6**

PROJECTS OF LITTLE INTEREST

	Year 1	Year 2+	Total
Total	112.3	236.0	**348.3**

FIGURE 17.1 The sales funnel.

A brief explanation of the amounts shown follows, through some considerations:

- Let us assume that it is the beginning of year 1.
- The sales organization is aware of a number of projects, for an overall estimated value of 630.9 MMUSD.
- Part of these projects is of interest, part is not, as it is normally believed that there are no chances on them. Why to keep them in the database? – may somebody ask. The reason is that it is always good to have knowledge of projects, for many reasons, from statistical to workload projection of the competitors.
- Specifically, there are 282.6 MM USD worth of projects that display an interest (from high to low) and 348.3 MM USD in projects of no interest.
- These projects are allocated in time, using information as well as assumptions that may be subjective but are much better than nothing.
- The high-probability projects represent less money than the others. Limited to the projects of interest for year 1, the high ones represent 18.0 out of 208.8 MM USD, are therefore around 9%.
- This percentage is even lower if we analyze the farther projects (Year 2+), where high projects represent 5.0 out of 73.8 MM USD, or less than 7%. The difference is not huge, but significant.
- The projects that are farther in time (Year 2+) are generally less probable.
- The funnel clearly shows the effects of the work for continuously improving chances, reducing the company's target from a broad all-inclusive mass of projects to a few highly probable ones.

SALES PEOPLE

In order to understand "who" the sales people are, a concept that differentiates project-oriented companies from product manufacturers should first be identified. In the latter, "sales" is generally the function that is responsible for following up a market, staying in touch with the customer and obtain the order. In these organizations, proposals have a role that is generally relegated to preparing a cost estimate. On the other hand, in engineering and construction contractors, the responsibility of the proposal team is much wider, because their proposals contain elements of technical, organizational, project management, scheduling, commercial, cost estimating, and pricing. As it will be mentioned in below paragraphs, a proposal is often a mini-project and so the responsibilities and the skills of those who have to prepare them are much more complex.

Having said this, in complex worldwide organizations there are two types of sales people, who coexist and support each other. On one side are the territorial sales persons, on the other the technical sales persons.

Territorial sales people are those who reside in a given country or visit it very often and are responsible for establishing and maintaining a relationship with the potential customers at every possible level. They are familiar with the top managers, as well as with middle management in charge of business units and even to technical people who are encouraged to share their technical problems, so that the sales person can facilitate a contact for a technical approach.

Technical sales persons are the experts of the products to be sold, sometimes known as product managers and they are able to effectively promote their product with the customers. This provides the condition of becoming aware of the problem that a customer may have and get a chance to propose a solution that solves the problem. Their involvement with a customer may be the outcome of a regular schedule of visits, or a call for support by the territorial sales person. In either case, it is essential that technical and territorial people act with the tightest coordination, to show the customer that "we are a team". Cases in which technical people show up in front of a customer without the person in charge of the area being informed, unfortunately occur and give a very bad impression to the customer: "If you are not able to coordinate between you, how can I hope that you will work to give me the best solution?"

Generally, territorial people are the "owners" of potential projects in the regions or areas assigned; among their tasks, they are responsible for the maintenance of the sales funnel. Product managers are also in charge of the development of the product and the technologies in order to maintain and enhance their characteristics of validity and competitiveness that are essential for selling it.

There is a third type of sales people, which is typical of large, global contractors who deal with large global customers. To do so in an effective manner, it is not enough to meet and follow up with regional organizations of the customer, but another function is needed with the global responsibility of interfacing the same customer wherever it is. This figure is generally known as "key account manager" and he/she must be able to direct the effort of both territorial and technical sales organizations in order to capture the best opportunities with that given customer, wherever they present themselves.

Needless to say, the introduction of a key account manager requires even more internal coordination, as the risk of an uncoordinated approach becomes higher, potentially spoiling the effort that the company is making to keep that given global customer under follow-up.

PROPOSAL PEOPLE

Proposal people are those who prepare tenders and bids to respond to invitations received from customers. Their activity generally stretches up to and including signature of a contract and even the start of project activities in coordination with the assigned project managers.

Proposal managers work on their proposals like project managers work on their projects, with the notable difference that, in an Engineering-Procurement-Construction (EPC) contractor a proposal lasts much less than a project. A typical Lump-Sum Turn-Key (LSTK) proposal can take 3 to 4 months in total for its preparation, depending upon its level of complexity. On top of this, the time for follow-up, negotiation and finalization is to be added. The overall effort may take 6 or say 9 months from inception to winning; after which the project manager who will inherit the project will be busy on it for the best part of approximately 3 years.

The set of activities for the proposal stage is almost similar to project execution in terms of headings:

- Process work to define the basics
- Detailed engineering work, however, only to the extent necessary for the level of estimating required
- Procurement and subcontracting work, to a similar extent as above (e.g., collecting bids for major/expensive items without any orders placed)
- Estimating work based upon an agreed plan
- Planning work
- Commercial and contract management work

Please refer to Appendix 1 – Cost Estimating for further details.

In this respect, it can easily be stated that a proposal is a mini-project, with almost similar activities performed in a much shorter period. Proposal people are usually the busiest in a contractor's organization, lights in their offices are often on until late hours, while most of other employees are sleeping in their beds.

The main differences between a proposal and a project are worth observing:

- The entire EPC implementation, which in a project takes normally about 2–3 years (depending on the project size and complexity); this is to be completed within a much shorter period for a proposal.
- The coordination requirements are therefore relatively tighter.
- The attention by the company senior management is higher.
- There is a deadline represented by proposal delivery, which cannot be missed. Of course, a project has time obligations as well, but there are remedies, such as Liquidated Damages and generally a day delay does not hurt. For a proposal, things are different: miss the date by a half day and you will miss the job. In the history of any contracting company, there are no examples of missed proposal delivery dates and if there is one, it is remembered as a remarkable disaster.

A proposal is eminently a multi-function effort. Every department of the company must contribute by taking up its portion based on its field of expertise of the proposal work in order to make the result attractive to the customer and maximize company's chances of success. The proposal manager is responsible for this collective effort and must be able to have a word on each of the many technical or commercial issues that constitute the proposal.

For small jobs, however, often there is neither time nor budget to involve all the departments and the proposal team becomes leaner. In some cases, the proposal manager is the only person who works on a proposal as a multi-tasking manager conceiving and writing the entire proposal.

Drawing a profile of a proposal manager is not an easy task, as organizational concepts, roles, and responsibilities vary greatly from a company to another. To exemplify this concept, Fester Consulting has made a survey on the recruitment advertisements published by a number of international companies in the mid-2010s. Specifically, two elements contained in the advertisements were analyzed:

- The responsibilities associated with the position
- The skills required of the candidates

The advertisements considered are from 16 different companies worldwide. The results have been plotted in the two figures that follow (Figures 17.2 and 17.3), showing the number of times that a certain element is mentioned in the job description. So, for example, the element "coordinate technical proposal team" is mentioned 13 times out of 16 among the responsibilities, while "participate in the BNB" (Bid-No-Bid decision) is present only once.

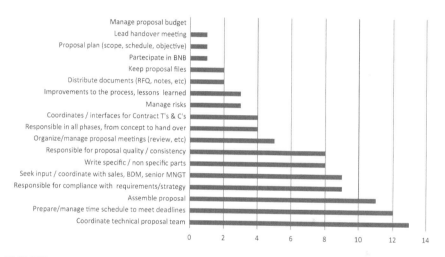

FIGURE 17.2 Responsibilities of a proposal manager.

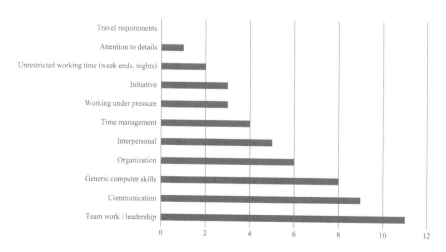

FIGURE 17.3 Skills required of a proposal manager.

The results are apparently surprising but need to be interpreted in the light of the organizational concepts prevailing in most companies advertising this type of position. Typically, the proposal managers have a boss who in some organizations is called a proposal director, in others a manager of proposals, or a business manager. This position reports to the commercial director and is the real person in charge of the proposal. The proposal manager is often viewed as an executor, or a coordinator. This somewhat longer hierarchical chain in proposals is justified by the very high attention by the senior management, as wrong decisions in this phase may have very negative impact on the project, the company and often even its survival. Therefore, it makes perfect sense that the proposal management work be split between two levels. One dealing with strategic and purely commercial matters, including costs and pricing. The second is more concentrated on the internal front to make sure that the final result is the outcome of a real teamwork among the various components of the company.

Similar considerations can be made by looking at the distribution of the skill requirements, where we see the prevalence of team-oriented characteristics like organization, teamwork, communications, etc.

INTERACTION BETWEEN COMMERCIAL FUNCTIONS

The complexity of the systems described above is such that the functions involved in the job acquisition process – sales and proposals – have many interfaces to be managed. Needless to say, the driver of such interaction must never be internal competition or rivalry. It is only through coordinated effort by all components that the job acquisition process can be successful, securing good jobs to the company. The following lines will explain how the two systems interact in the various phases of the process.

Once a prospective project materializes into a request for proposal, the sales organization hands it over to the proposal group. Here, it is necessary to determine more in detail if the proposal group can process the prospect. This phase is an "intelligent" handover, i.e., not just a transfer of documents, as the sales group should prepare a strategy document that explains how to do the project, presenting the chances, the opportunities, the risks, and so on. Obviously, long-awaited and well-known RFPs are immediately accepted, but the screening applies in general and it is a very important step, because the proposal group must make the first very important decision, called Bid-No-Bid decision, often referred to as BNB. Many companies have structured processes for this step, the purpose of which is to understand if the prospect is compatible with the priorities, budget, and workload of the proposal group and the company in general. A proposal manager carries out this process and this is generally the first time he or she becomes aware of the details of the job. It is also the first instance in which the two groups come to interact.

Once the decision to bid is made, the proposal preparation phase starts; at the onset, there is a handover step and the job is taken in charge by the proposal group, which will maintain responsibility over it until the conclusion, i.e., until the bid is won or lost. Such a responsibility includes the proposal manager acting as the official focal point for the communication with the customer, in line with the rules given by the RFP procedures.

During this long phase, the sales group does not disappear from the scene; on the contrary, it operates somewhat "behind the stage" in order to secure what is sometimes called an air cover. The sales group has the possibility to enhance the communication with the owner through channels different from the official one and so managing messages that could hardly be exchanged in an official way.

If the proposal raises the customer's interest, the contractor may then enter a phase of finalization and contract negotiation. This phase usually sees the proposal manager leading the effort of a team that includes technical engineers, project managers, and contract managers. The presence of the project manager in this phase is very important, as he or she will have to manage the job in case of success and so, attending to the finalization phase gives him or her the feeling and the knowledge that will be of an invaluable importance for a smooth and successful execution. Whoever has participated in the final negotiation of a contract knows very well the importance of the "air cover" in this phase, too. Also, in this phase there are messages that must be passed unofficially. Sometimes, to overcome a stumbling block, an explanation in an unofficial way, behind the scenes can be extremely useful to prepare the ground for a smooth final discussion.

The discussion on job acquisition process would not be complete if the contract handover step is not referred to. This process requires the proposal people to transfer to the execution people a whole set of documents that will be the basis for the next phase. Main documents are the contract itself and the budget, which are both generated by the commercial group and constitute the basis of the project management work. However, the transfer of key documents is just a part of the handover process. The proposal manager has followed the project from the inception phase to the contract finalization and knows every detail of it, including gray areas, feelings, threats, and opportunities. It is, therefore, fundamentally important that these evaluations be shared with the execution team. A good handover phase will ensure proper attention to those points that could pose risks or provide opportunities for extra earnings or benefit to the company. Impressions can include immaterial evaluations on the client's personnel, including anticipated problems with particularly "difficult" persons and it is always good for the project manager to be aware of such problems in advance. Ideally, the handover should be carried out based on a procedure and a checklist that should be part of the company's quality manual. In addition, minutes of the handover meeting should be prepared and kept for reference.

A good handover process is essential for an effective contract management, because the early identification of opportunities and risk could be the basis for the setup of claim or claim defense avoiding unpleasant surprises later on.

18 Contract Administration

Of all disciplines contemplated in the project management world, contract administration is the youngest and it is not yet widespread in the contracting community. In reality, the above statement is not entirely correct, if one looks at the content and the activities that fall under this discipline. In fact, the various activities included in the contract administration scope of work have always been there and somebody has always been doing them. The big innovation of this novel discipline is that it groups various activities and "somebody" who deals with them has now a precise identification, as the contract administrator.

The doubtless advantage of this novel arrangement is a more focused approach to the administrative and contractual aspects of project management, which are becoming more and more demanding in such a time when contracts are becoming every day more price competitive and time demanding.

In modern days, contracting in all sectors is affected by the specter of cost reduction, mostly due to shrinking of operating margins for production and manufacturing companies. This in turn forces operators to seek price reduction of their outsourced services, triggering a chain reaction, where everybody is seeking cost reduction in order to survive with decreasing selling prices.

Similar constraints apply to the ever-tighter delivery terms required for new contracts, resulting in further increase of project risk profiles, as liquidated damages may have a fatal effect on the already difficult financial conditions.

Under these predicaments, it becomes natural to look for every possible means to round up such tiny margins, or at least to protect them, lest the contract economic result ends up in the deep red. Hence, the necessity to consider claims and to defend against possible passive claims. There is broad concurrence nowadays that a good system of contract management represents one of the most effective means to achieve this.

A deep knowledge and an accurate management of all contract clauses and provisions allow both parties to ensure that the performance of the contract strictly adheres to the framework and the provisions negotiated. No matter who first introduced these techniques, nowadays both project owners and contractors make extensive use of contract management and administration as a body of knowledge incarnated in a well-defined professional profile.

Contract management services are often provided by specialized consultants putting their wealth of experience and record of accomplishment at the services of the company in need. This kind of services have many facets, from formation and training to coaching and training-on-the-job of company's executive and employees, up to temporary management, i.e., assigning expert contract managers to work inside the company's structure. This scheme allows working side by side with the project manager and his/her team to advise and act in order to manage key contractual aspects in a professional and consistent manner. As a longer-term result, the team will grow and develop this kind of professionalism for future projects.

Using these – indeed not cheap – services in times of crisis might sound like a paradox, but in times of crisis no stone must be left unturned when the survival of the company is at stake.

HISTORICAL BACKGROUND

To understand the background of contract administration and the history of its introduction and development in the contracting industry, one must refer to the different approach of contract economics by the project owner and the contractor. Generally, the project owner enters into a contract in order to implement something – an installation, an infrastructure, a set of services – finalized at a long-term production of goods and services. It ensues that the economic perspective of project owners is generally a long-term one. In a complex financial model, the variables are so numerous that often there is room for a recovery of margins also in the presence of higher-than-planned investment costs. It does not mean that the project owner's financial model is not sensitive to the investment cost, but it should not be forgotten that the investment cost is only one of the economic factors that determine the viability and the good economic result of an initiative. In some cases, raw material price or final products sales price can be key to the success – or lack thereof – of an industrial initiative. In some cases, the Operating Costs (OPEX) represents the key factor in the financial model, while in some others the Capital Expenditure (CAPEX) plays the main role in the model.

The case of the contractor is completely different, as the economic and financial result of the contract represents 100% of the eggs, which are put in the same basket. There is no way to cope with a negative result deriving from a cost overrun or the application of liquidated damages and the figures in the red (negative) in a big project's budget are often the prelude to red results in the company's financial statement for one or more years.

If one thinks of this difference, it is not surprising that contract administration was firstly introduced by contractors to protect their interest and in some cases, their own survival. The main objectives of strengthening this area were as follows:

- Better identification and management of change orders, as money was often spent in supplies that were not included in the contract and came as the result of requests from the owner's technicians, not effectively managed by the contractor's project management team.
- Follow-up of counterpart's obligations, as sometimes a project can go wrong for the contractor because the owner fails or delays the performance of some of its obligations, which are vital for contractor.
- Identification of claims and management of dispute processes. In times when dispute resolution was not a well-developed knowledge, the best way to "win" a claim was to strike first and strike hard, with a good analysis of contract provisions and a punctual identification of claim items.

For these reasons, the initial focus of contract administration was on claim management on contractor's side and then all administrative aspects became part of the typical set of activities the new discipline would deal with.

Subsequently, project owners started seeing the usefulness of a good structured contract administration and large companies – especially those of Anglo-Saxon origin – have introduced it massively.

In present times, it is not uncommon to see large Engineering–Procurement–Construction (EPC) contractors and global oil companies (and the like) with a contract administration department staffed with 100 or 200 engineers and lawyers.

Unlike other project management disciplines, the definition of contract administration is still vague and the duties and responsibility of the professionals in this area are not the same in every company. Likewise, there is no full concurrence on the reporting lines both at project level and at corporate level; different schemes and approaches are adopted by different companies. These anomalies are a clear indication that this is a young discipline, the outline and details of which are still being defined in the world of project management. There are signals that the industry is evolving toward solutions with the highest value for money. In some cases, areas that were followed by other people in the project team are little by little attracted inside the contract administration area. In this respect, the evolution of this discipline is anomalous, as the debate is not about things to do or how to do them, but it is primarily about who does them and to whom this position reports.

In the following paragraphs of this chapter, a broad scope is contemplated as opposed to a narrow one as preferred by some authors. The main areas which will be referred are as follows:

- Administrative aspects
- Contract management
- Change management
- Claim management
- Business aspects

ADMINISTRATIVE ASPECTS

This is the first and perhaps the most widely accepted area of action of a contract administrator, covering the purely administrative aspects. Very importantly, the contract administrator is active through all the phases of a project, including the pre-award phase, i.e., proposal, negotiation, and contract formation. The scope of work of a contract administrator is described next in the form of a job description.

PRE-AWARD

- Familiarize with contractual documents.
- Give recommendations as to price formation.
- Assist preparation of initial contract document, or give comments to the text provided by the owner in the Request for Quotation (RFQ).
- Participate in the risk analysis and setup of risk management process.

Post award – Setup

- Familiarize with the following:
  - Project procedures, both external and internal
  - Management tools such as SAP or similar
  - Contract price, price schedule, terms of payment
- Update or ensure the update of the following:
  - Valid insurance certificates
  - Bank guarantees
  - Time schedule, including keeping track of previous versions
- Highlight areas of concern, risk or opportunity in the contract.
- Support the project manager in determining the contract administration requirements and ensure that the function is adequately staffed at project team level.

Correspondence

- Ensure that proper procedures are in place and are correctly followed for the following:
  - Filing of incoming and outgoing correspondence
  - Updating correspondence log
  - Monitoring unanswered correspondence

Meetings

- Handover meeting
  This is the internal meeting in which the project is handed over by the commercial team (proposal manager if on contractor side, or buyer if on client's side) to the project team (project manager).
  - Ensure that a handover checklist is in place.
  - Make sure that project documentation is properly transferred and is in the right revision.
  - Ensure that project challenges, risks, and opportunities are adequately understood.
  - Make sure that gray areas are covered, both as risks and opportunities and that focus will be kept on those areas throughout the project execution.
- Kickoff meeting
  It is the meeting that takes place after contract signature between owner and contractor, in which detailed project and contract aspects, for which there was no time prior to contract award, are discussed.
  - The agenda is mostly dictated by the contract and comprises subjects the definition of which was not possible to achieve during contract negotiation and was postponed by agreement between the parties.
  - The meeting must be held soon enough after the contract effective date (start of the works) in order to define important pending matters, but not too soon (e.g., few days), to give the Parties the time to prepare the meeting adequately, including preparation of documents to be handed over at the Kick-Off Meeting (often the "KOM").

- Periodic project review meetings
   These are the meetings between the parties intended to monitor the correct progress of the project works.
   - Monthly progress review meeting, where the monthly progress report is discussed
   - Weekly change order review meetings, when justified by the size or the complexity of the project
   - Weekly construction progress meetings at the construction site, to review progress, anticipate issues, and plan ahead
   - Daily construction meetings at the construction site, to review the day's planning
   - Daily safety review meetings at construction site, because there is nothing more important than safety

PAYMENT REQUESTS AND INVOICING

- Ensure that the milestones for which a payment will be requested were properly achieved and the achievement is properly documented.
- Ensure that support documents are properly prepared, in order to avoid the risk of rejection with subsequent payment delay.
- In particular, ensure that the payment certificates include all necessary elements such as retentions, deductions, back charges, and the like.
- Once the payment requests satisfy the contractual requirements, release them for official transmission to the owner.

CONTRACT MANAGEMENT

A contract is often a sort of jungle of documents, annexes, appendices, attachments, exhibits, and similar. To make things more complicated, each of these documents can in turn have its annexes, appendices, etc. The complication reaches a climax if these documents have different origin, such as licensor's packages, vendor's documentation, and standard documents. In these cases, the terminology can be different so that in one case we can have an annex to an appendix, in others there are appendices to annexes and so on.

Another aspect that makes contract management a complicated function is the order of precedence among all the documents forming part of it. This is essential in order to settle possible conflicts among different parts of the contract, which – inevitably so – impose inconsistent requirements on the same subject.

The contract administrator must have the ability to analyze and suggest positions on various issues such as:

- Whether a request from owner is a contract change or not
- Whether a claim or a request by the owner is grounded or can be rejected
- Whether there is ground for a claim to the owner in respect of its incompliance with contractual obligations

Therefore, contract interpretation is one of the key tasks of a contract administrator, who is often requested by the project manager to advise a position in respect of a controversial issue. Needless to say, things are not always black and white and almost every question submitted to the contract administrator has a range of answers in shades of gray, usually not black or white. In many cases, support by the legal department is also necessary, for those aspects that have legal significance.

The final decision is left with the project manager, for example, whether or not to file a claim, whether or not to refute a position by the owner, etc. There is a moral in this story: Decisions are always a management issue, but a good contract administration process helps the management to make informed decisions.

There are other ways in which the contract administrator can support the project manager and the project team in respect of the contract text:

- Support preparation of contract amendments, revisions, etc.
- Keep an updated version of the contract and its various annexes, incorporating all revisions and amendments. This will be a sort of prêt à porter master copy for easier consultation.
- Prepare a synopsis of various key contractual aspects, with a summary of the applicable provisions and a cross-reference to the relevant contractual points.
- Give periodic briefings to the project team on specific aspects of contractual significance, in order to spread and enhance the contract awareness in the team.

One point must be clear when discussing contract management and relevant responsibilities: the final accountability for the contract is and remains with the project manager. He/she will never be relieved of the responsibility to know, understand and manage his/her contract, just because a good contract administrator is able to answer any question on it.

CHANGE MANAGEMENT

The first important principle is that a contract is not allowed to be modified without the approval of the authority who has signed it, be it the CEO, or managing director, or whatever authority is designated to do so by the company's procedures. This provision must be present in the contract and is essential to prevent undue modifications of arrangements that have been negotiated and approved at high corporate levels. Uncontrolled modifications would mean anarchy and the contract would not be manageable.

On the other hand, some flexibility must be left to the project managers, in order to avoid recourse to the senior management for every minor issue that requires fine-tuning of contract provisions. Therefore, the contracting theory has introduced the concept of variation orders, or change orders. These are minor modifications of scope that can be discussed and approved at project level, so that the authorization by the project manager is sufficient for it to take effect and have contractual validity.

CONDITIONS FOR A CHANGE ORDER

Some conditions should exist, so that a change can fall into this category and higher level approval is not required:

- The nature of the additional work must be the same as the main nature of the contractual works. For example, in a contract for the construction of a hospital, installation of an oxygen distribution system, which is inherent to the nature of the hospital, can be introduced. On the other hand, the construction of a shopping mall on an adjacent area would be something of a different nature, even though the owner may be the same. This does not mean that it cannot be done; it only requires a formal modification to the contract, in the form of a contract amendment, approved by the same authority that has approved the original contract.
- The amount must be minor, i.e., it should not bring in a substantial increase or decrease in the contract price. In some contracts, in particular for public works, a limit is pre-set and for instance, 25% may not be an unusual figure.
- The owner should have enough budget allowance to fund the change. If the budget is not enough to cover for the variation, then a process must be activated inside owner's organization in order to provide the necessary additional budget; such process must necessarily end with the approval of the high-level authority that has approved the initial allocation.

Once again, these conditions do not mean that the change cannot be approved and implemented; it simply means that approval cannot be released by the project manager.

CONTRACT AMENDMENTS

Modifications to the contract, implying higher-level approval, are known as contract amendments and they have high contractual priority as they have been expressly negotiated to modify and supersede certain contractual provisions.

Negotiation of a contract amendment can be carried out at project level, sometimes involving the legal department and certainly involving the contract administrator, if such position exists in the project.

CHANGE ORDER APPROVAL

The procedure for processing and approving a change order must be specified in the contract, sometimes in the main body, in some other cases, in an attachment. The main principle is that changes are generally requested and ordered by the owner, because the owner considers the modification necessary or beneficial for a number of reasons, amongst which are the economic performance of the plant, or improved maintainability, operability, and the like. Please note that a change order is also named as an extra work request, or extra work order.

In the most general case, the owner representative would instruct the contractor to modify some element of the project, either by adding something, or by modifying something already completed or being completed; the contract generally contains a provision stating that such instructions must be followed, irrespective whether they imply cost increase or additional execution time. Of course, in case the contractor considers that the instruction would cost him money and / or time, it would point it out officially, informing the owner that the instruction is a change order.

The owner at this point can (i) confirm the instruction stating that no compensation is foreseen because the solution was already included in the contract, or (ii) confirm the instruction, requesting the contractor to quantify the extra cost/time, or (iii) withdraw the instruction.

In the first case, the contractor is obliged to execute the change, unless there are safety reasons preventing it. At the same time, the contractor would raise a claim, reserving to prove the nature of a change and quantify the relevant impact.

In the second case, the contractor would prepare a proposal for the change order, giving details of the technical solution, a cost estimate and an evaluation of additional time required, if any. Once the owner is in receipt of the change proposal, it can (i) approve it, in which case a change order will be drawn up and signed by the project managers, or (ii) approve the execution but not the price, in which case it will instruct the contractor to proceed with the modification and submit evidence of the costs borne, for reimbursement by the owner, or (iii) cancel the instruction.

The above is a typical scheme for processing a change order; it is interesting to note that contractors are generally not enthusiastic about change orders, as they distract the focus of the project team from the mainstream execution to the preparation and discussion of the change order. One deterrent effect that the contractor often has is the provision that entitles contractor to the reimbursement of the costs borne for the preparation of the proposal, in case owner decides not to proceed. Generally, the amounts of money involved are minor, but having to approve and bear small additional disbursements is annoying for the owner and it is a good barrier against uncontrolled change order requests.

The above approval procedure can be seen in a flow chart form in Figure 18.1.

ROLE OF THE CONTRACT ADMINISTRATOR FOR CHANGE ORDERS

The first and outmost decision when the owner gives an instruction to modify something is answering: "Is it a variation to the contract?" The answer to this question is sometimes straight forward, but in some cases, it requires a thorough analysis. This exercise is a typical task of the contract administrator's and sometimes it requires even the support of lawyers.

Unfortunately, notwithstanding the accurateness and the depth of the analysis, it is not always possible to obtain a clear-cut answer and once again, the project manager will have to make a managerial decision.

Another important role of the contract administration in respect of change orders is to keep a log of all the cases, categorizing them as approved, withdrawn, pending, and rejected, for the sake of good management practices, as well as for monthly progress reporting by the Project Manager.

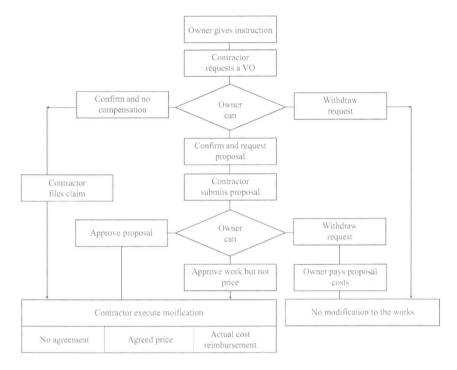

FIGURE 18.1 Typical approval cycle for a change order.

CLAIM MANAGEMENT

A claim is something completely different from a change order. This statement is perhaps obvious, but in too many cases, the two concepts, which are in fact quite different, may easily be mixed up by some project team members.

A claim is the entitlement – or deemed entitlement – to compensation for events that in the opinion of the claiming Party have altered the original contractual picture. Scope of this chapter is not to enter into the details of claim strategy, which is outside of the scope of this book. The focus will rather be on the nature of a claim, the attitude of the parties, and once again, the role of the contract administration function.

There may be many reasons for raising claims and there are contractors who regard them as a way to enhance project margins, which are becoming tighter and tighter because of the growing competitiveness of the marketplace. In fact, from a purely budgetary perspective, money received in respect of a claim is pure additional money for the project's accounts.

Generally, a claim can be rejected or accepted and in the latter case it will give origin to a change order. If rejected, it will be recorded as a claim and be handled according to contractual claim procedure.

Another very important principle is that contractor should never suspend the works if owner rejects a claim. Such a negative behavior, besides going against specific contract provisions, is generally regarded as lack of seriousness and can cause loss of reputation to the contractor. Sometimes the damage arising from this attitude can be much worse than the damage caused by the owner's refusal to accept the entitlement, particularly in closed markets, where everybody knows everything of every project.

EXAMPLES OF CLAIM

A claim can occur in any phase of the project, although it is mostly in the engineering and construction phases. As to give some typical examples, contractor raises a claim because one of the following events occur, and the owner refuses to acknowledge this as a change order:

- Owner has requested to modify the layout of a unit, altering the one that was defined in the contractual documents.
- Owner has requested to install a spare unit on a pump that had no spare in the contractual documents.
- Some of the codes and standards indicated by the owner in the RFQ and included in the contract, are found to be not applicable.
- Codes for seismic design have changed in the country where the plant is to be built.
- Owner has requested a higher corrosion allowance for materials in a given section of the plant.
- Owner has requested too many change orders; although they have been paid, still there is a disruptive effect because of massive redoing of documentation already prepared and approved.
- Owner orders the reopening of an excavation already backfilled, to carry out an inspection of an underground system.
- The pipe rack needs to be expanded because owner has requested to install bypass lines for certain equipment.
- Owner has requested to uninstall valves already installed and move them half a meter down, because they are too high for the operators to turn the handwheel safely, as the average stature of operators is low.
- Owner's supervisors are trying to interact directly with construction subcontractors, giving direct instructions that conflict with contractor's instructions (management interference).

The above are examples, but the cases shown are not unrealistic at all, as situations like those happen normally in the life of a project. If we analyze each of the above occurrences closely, we can note that the situation is never straightforward.

As a matter of example, let us take the case of the valves to be relocated. Contractor will claim that the drawings and general specs provided by the owner with the RFQ and included in the contract, made no mention of such a requirement. So, contractor met the requirement of handwheel elevation specified in the contract; therefore, the issue is new and the owner should bear the extra costs and grant time extension. The owner will maintain that the contract states that the plant shall be designed in a manner to be safe, operable and maintainable and this general requisite supersedes the details of the drawings, which are to be considered for information only. In addition, the contractor had to obtain information about the stature of owner's operators before confirming the elevation of handwheels. As we can see, the discussion will be endless and so the contractor will execute the modification, but claim an entitlement to compensation for cost and extra time.

TIMING FOR CLAIM PROCESSING

By its nature, a claim must be raised as soon as the claimant becomes aware of the event. Often, contract s state a dead line for the notification, after which the entitlement is lost. This is a very serious matter and the contract administrator must be very well aware of it when planning to raise a claim.

Discussion and finalization of a claim can happen immediately, or it can be postponed to the end of the project. There are pros and cons in the postponement, the main consideration being that contractor holds the knife's handle as long as the plant is under construction. Once the plant is built and delivered, the handle turns by 180°, the owner will hold it and the negotiation power will swing toward the owner. On the other hand, a hasty conclusion of a claim can sometimes be detrimental for contractor, if the solution is tombstone-like, i.e., it will preclude any rights to reopen the discussion for events having an origin before the settlement of the claim. In this way, new consequences of old problems will be denied the entitlement to compensation.

ONSET OF CLAIMS

Since the time of proposal preparation and contract negotiation, skilled proposal managers, and contract administrators can spot risks and opportunities in the pages of the contract and in particular in its gray areas. It has already been noted that one of the purposes of the internal handover meeting is the identification of these gray areas. Because of that, the project manager knows that certain areas are to be given particular attention during the life of the project, as eventually a bell can start ringing for a real opportunity and it can be realized that the ground for a claim is becoming solid in the short term. Therefore, it is a good practice that each piece of documentation, including minutes of meeting, specifications, correspondence, etc., relevant to such gray areas, are duly tagged and kept available for use as a claim. This task is on account of the contract administrator, who will continuously sensitize the project team on these areas, so that whenever a discussion starts on these subjects, the level of attention immediately escalates.

The main purpose of claim anticipation is to have the documentation readily available. A good claim submitted at the proper time, supported by all the relevant documentation is somewhat difficult to refute. With a soccer-based metaphor, we can say that in the World Cup final the team who scores the first goal has good chances to win and raise the cup to the joy of its supporters. This does not necessarily mean that the team shall win, but a substantial step toward victory may have been taken.

CLAIM MANAGEMENT

A claim can be represented as a three-legged stool. Obviously, the stool cannot stand on its own if one of the three legs is missing: It will miserably fall down and everybody will look at it and say "what a useless stool". Likewise, a claim must have three components:

- A documented event
- A contractual basis
- A notification in accordance with the agreed procedures

If one of the legs is missing, the claim will tumble down like the poor crippled stool and everybody, mainly the lawyers of the counterpart, will say "what a useless claim".

Technically speaking, the documented event and the contractual basis are called the "an". In Latin, an means "whether"; whether or not there is basis for a claim. Once the an is ascertained, it will be necessary to concentrate on the "quantum", i.e., the quantification of the entitlement, which will have to be provided in accordance with the agreed procedures, in terms of timing, requisites, and back up information.

A claim is a very rational exercise and both the an and the quantum must be carefully analyzed and even challenged internally, in order to make sure that once the claim is submitted it will be solid and defendable. Engineers as well as lawyers have a rational mentality and they know the importance of the cause-effect relation. The contract administrator, or in this case the claim manager, is a professional figure that has part of the competences of an engineer and part of the competences of a lawyer and is best indicated to bring about the claim with the best balance among its three legs. If the an and/or the quantum are not defendable, then the situation will be the same as the soccer team who is given the opportunity to score first in the World Cup final, but it misses the penalty: Many will see it doomed in view of the final result.

Generally, the level of attention of a claim is very high inside the company, because a dispute is something that can touch very sensitive strings, involving the strategic position of the company in the market place, which can be affected by a perception of excessive contentiousness. In addition, relationship with the Client can be spoiled by big claims, especially if they end up in court or in an arbitration tribunal. This explains why the contractor's senior management is usually very interested in the development of claims and follows it carefully, through the internal counsel and in many cases through external counsels as well.

ROLE OF THE CONTRACT ADMINISTRATION

The role played by the contract administrator in a dispute is very important as the bridge between the project team and the counsels, both internal and external. The particular "sub-species" of contract administrator who is involved in this role is a claim manager, also sometimes referred to as a "legal engineer".

Depending on the actual organization of each company, contract administration can deal with active claims and/ or passive claims, i.e., claims against clients or against subcontractors. In some companies, the contract administration group – or department – includes both the administrators of active contracts and of subcontracts. There are also many examples of contractors – and they are probably the majority – where active and passive contracts are managed by two different groups and the subcontract administrator would report to the construction organization. Needless to say, both active and passive claim management include claim and claim defense, i.e., the activities necessary to prepare, submit, and manage a claim to the counterpart, as well as those necessary to defend against a claim submitted by the counterpart.

DISPUTE RESOLUTION

The contract must be very clear as to the mechanisms for dispute resolution. Generally, there are two levels: a first level consists of attempting an amicable resolution. Failing this, there is generally a second level that consists of courts or tribunals. There are important differences between dispute resolution in courts of law (ordinary justice) and in arbitral tribunals and either solution has pros and cons.

Seeking resolution in a court of law has the main disadvantage of longer and somewhat unpredictable timeframe, also based on the actual country where the court is located. In almost all national systems, the judicial systems have two or three degrees of judgment, as a ruling can be challenged in one or two stages of appeal.

Arbitral tribunals provide higher certainty of the timeframe and the ruling (technically, an "award") cannot be challenged by either party, because they had previously agreed in the contract's arbitration clause that the award should be final and binding. On the other hand, the cost associated with arbitration is generally higher. Notwithstanding this, arbitration is becoming more and more popular and most international arbitration organizations are making efforts to amend their procedures in order to make it a more flexible instrument and reduce time and costs.

Another important factor is the enforcement of the award. In the case of ordinary justice, the enforcement is automatic and not questionable, as the judge would issue an executive order for the compensation awarded. The convention of New York, which most states have ratified, makes the enforcement of an arbitration award mandatory. However, as the enforcement will often take place in the owner's country, it will be subject to that country's rules and regulations. In many cases, failure to comply with a specific form prescribed by local legislations can even nullify the effect of the award. Counsels and arbitrators should take particular care in making sure that the award is exempt from defects of form that could cause its non-enforceability in the country.

BUSINESS ASPECTS

Another important area of action of contract administration is the management of the business interfaces of the project team inside the company. There are a number of dedicated company's departments covering specific areas of business, with which the project team needs to have an operating interface for the carrying out of project activities, like for instance,

- Financial, for payment instruments, letters of credit, invoicing procedure, bank guarantees, etc.
- Legal for contract interpretation, amendments, claims

- Insurance for policies and claims
- Human resources for assignment conditions of personnel abroad
- Local organizations for taxes, social security, etc.

The function of the contract administrator in this area is to manage these interfaces, on behalf of the project manager. As a matter of example, financial department has the institutional role to manage external interfaces such as banks, but contract administration will manage the interface between financial and the project team.

This particular area is not always included among the tasks of contract administration and many companies exclude it from its duties. One reason is probably that it is not easy to find all the necessary competencies in one single person, but experience has shown that, when this happens, there are doubtless advantages.

ORGANIZATIONAL MODELS

As it is discussed earlier, contract administration is rather a young discipline and as such, its fields of activity, as well as the organizational models, are not universally recognized in the community of contractors. For the purpose of discussion, it is assumed here that there will be a central administration unit or function, which is responsible for the development and the maintenance of the specific know how. The same unit is also responsible for the effective and efficient provision of services to the divisions of the company that require them.

There are two questions to be answered when trying to define organizational schemes for contract administration: (1) What is the position of the central unit in the corporate organization? and (2) How does it operate in the projects?

The first question has different possible answers, each one with pros and cons:

- In the commercial division: The advantage is to optimize the involvement of contract administration in the projects since the pre-award phase.
- In the legal department: Interesting solution because of the many interfaces and similarities between contract administration and legal. The disadvantage is that contract administration needs to be active at the operating level, while the legal department, in general, has more the mindset of a counsel.
- In the project execution division: This appears to facilitate the direct contact and operating utilization of the function in the projects, removing problems such as definition of priorities among projects for the allocation of resources.

The second question – how does contract administration operate in projects? – cannot have a single answer, the most logical one being "it depends on the project". Assuming in any case a matrix organization, we can imagine a different approach toward projects based on their complexity:

- Complex projects, i.e., those where one or more of the following factors are present: Large economic value, multiple interfaces, complex financing scheme, high risks for liability, or other. For these projects, a typical matrix organization can be used, with one or more contract administrators assigned by the central function to the project team on full time basis for the duration required, often using a task force scheme.

- Projects having intermediate complexity, i.e., those of which economic size do not justify/afford the full-time assignment of a team. In this case, a solution could be to allocate some of the contract administration tasks to one or more employees in the project team. For example, project engineers can be entrusted with the monitoring of the change orders or the management of contract. These personnel can undergo specific training, at care of the contract administration department and work under an indirect coordination by the same. It is also possible to use personnel from the central function, without moving them into the task force, but simply on part time basis, or as needed basis.
- Simple projects, for which the contract administration services are provided by project team, or even by the project manager himself, in the same way that was used before the introduction of a structured central function. Some specific services can also be provided out of the central unit because they require specific competencies such as claim management or advanced contract management. These services can be provided on spot basis, i.e., as-needed basis.

As it can be seen from these cases, the most important thing is that the extent of the services must be agreed beforehand between the central unit and the project manager. Moreover, such an agreement must be made before the project starts and even before the proposal is submitted, because the extent of the services provided will determine the cost of this activity to the project.

PROFESSIONAL PROFILE

Although the discipline is relatively recent, there are several international associations that group contract administrators and represent a professional family. Among other activities, most of these associations organize meetings, seminars and conferences around the world, though they are mostly centered in the United States. In a typical year, it is not uncommon to have from 40 to 50 such events organized. The main purpose of these meetings is to allow people to share experience and to increase awareness of the role and knowledge of the techniques.

It is interesting to analyze a large sample of contract and commercial managers in order to understand who they are, what their job is and in what type of industry they render their services. To this effect, some interesting data were gathered in the mid-2010s on a sample of over 40,000 individuals and the following interesting conclusions were drawn:

- Figure 18.2 shows the distribution of the contract/commercial managers by their level in the organization to which they belong. Most of them are independent professional or middle managers, while few belong to the higher ranks of the organizations.

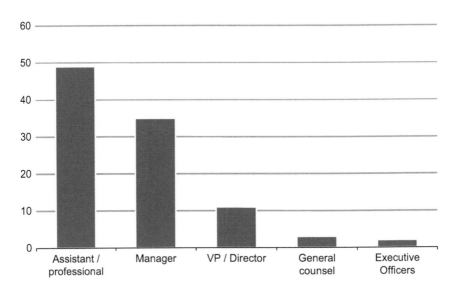

FIGURE 18.2 Distribution of contract/commercial managers by level.

- Figure 18.3 shows that most professional contract and commercial managers are working in commercial management or procurement organization. It is interesting though to note how non-negligible percentages of such professional are employed by the industry in other sectors, such as sales, operations, finance, and others. The reason is that the ability to deal with and manage a contract is important whenever a contract is involved and contracts are the basis of every industrial activity.

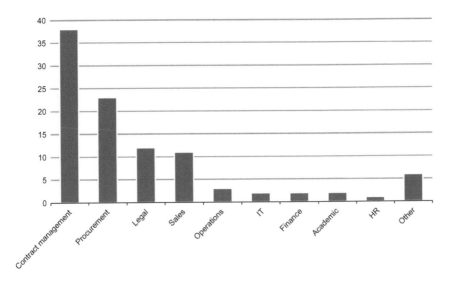

FIGURE 18.3 Distribution of contracts/commercial managers by function and role.

- Last but not least, Figure 18.4 shows the distribution by type of industry. Very interestingly, the distribution is rather homogeneous, showing that almost all sectors of the industry take benefit of the ability to manage contracts professionally. The reason why oil and gas takes the lion's share is probably that the concept of contract management and administration was born here. Among the other industries utilizing this category of employees, are the most technologically advanced sectors, traditionally more sensitive to innovation, like IT, consultancy, engineering, and construction. Other more "mature" sectors such as manufacturing and health care display lower utilization of this professional family.

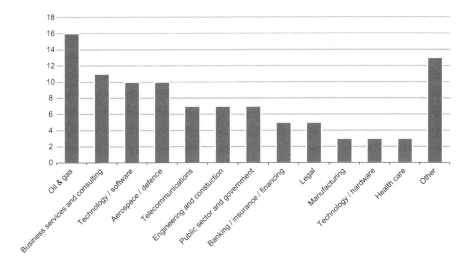

FIGURE 18.4   Distribution of contract/commercial managers by type of industry.

Appendix 1
Cost Estimating

Cost estimating is one of the most crucial disciplines for those who are involved in engineering and construction contracting, both from a project owner's and a contractor's perspective.

When a project owner studies a new project, it usually performs cost estimates at various stages of the definition process. The result of each stage of estimate will determine whether or not the project owner will move to the next phase. In particular, the outcome of the last stage of the process will release the funds for the implementation of the project. This is a major milestone for a project owner and implies sometimes a major allocation of funds; this shows the crucial importance of using an accurate cost estimating methodology.

Seen from a contractor's point of view, in the bidding phase the best proposal, outlining the best execution plan by using the best state of the art technologies can be prepared. However, if the cost estimate is not sufficiently accurate, two risks will immediately emerge. The first risk, dreaded by commercial directors, is an overestimate that would put the proposal on the high side. In this case, the customer will be compelled to discard it, with disappointment if it sees meanwhile that the technical proposal is quite good. If the consequence of an overestimate is losing a job, the effects of an underestimate can be much worse, causing the company to lose money by executing a lump-sum project at a price that does not cover the actual costs.

Reliable cost estimates are needed in every phase of a project, for example, in the definition of change orders for the implementation of variations to the original project scope.

The fateful effects of a wrong cost estimate may be much more worrisome if the reasons are structural, i.e., the procedures in use are inaccurate and tend to exaggerate or underestimate the costs. A good cost estimating manager should spot this risk and mandate periodic reviews of the procedures and standard criteria used, in order to come to a cutting-edge system, which is exactly what is needed. The main idea is that margins should not be built in the estimate at technical level; rather, they should be decided by the management. The cost estimate that is submitted to higher levels for the final proposal review should be bare and rigorously contain (and highlight) only those contingencies and margins that are foreseen by the applicable procedures. The final decision about contingencies, margins, and so on should be taken on the basis of commercial and risk/opportunity considerations. This is particularly true for the portion of cost estimate that is performed by discipline engineers instead of professional cost estimators. Specialist engineers systematically tend to overestimate by adding contingencies. The behavior is understandable because they are the ones who will ultimately have to manage the project within that cost and so they tend to make the budget "comfortable".

TABLE A1.1
Cost Estimating Classes

Estimate Class	Purpose	Project Phase	Typical Accuracy Level (±%)	Required Project Progress
5	First screening	Pre-feasibility, screening	30–50	2
4	Feasibility	Feasibility study	25–30	5
3	Budget	Basic engineering design	15–20	15
2	EPC tendering, budget control	FEED	10	30
1	Definitive estimate	EPC execution	5	80

ESTIMATING CLASSES VERSUS ACCURACY

Cost estimating can be a very simple or a very complex exercise and it determines the level of engineering required to support the estimate itself. For this reason, the type of estimate required depends on its purpose. International associations in the industry tend to use the concept of estimating classes, whereby each class has a purpose and the details about the procedures come as the consequence. Very importantly, each class should be linked with a different estimating methodology.

Table A1.1 shows a typical classification of cost estimates according to their purposes.

In normal practice, operators (both project owners and contractors) tend to refer to the accuracy rather than the estimating class. It is seldom heard a project manager say he/she needs a class II cost estimate, since the need is rather expressed as a ±10% or a ±30% cost estimate. Such a practice certainly lacks rigor, but the language has become universal and everybody says that for a LSTK proposal you will require a ±5% to ±10% accuracy, whilst some literature sources call for ±5%. This kind of approach can also be found in official documents like Requests for Quotation for basic engineering services, usually calling for a deliverable consisting of a cost estimate with a given accuracy.

Anyway, call it either estimating class or estimate accuracy, some criteria will be necessary. The procedures for each company will define the type of estimate necessary for the various activities and the purposes of such an exercise. So, a company's cost estimating manual may have chapters titled as "LSTK proposal", "Preliminary proposal", "Not binding estimate", "Twenty-four-hour estimate", etc., rather than references to classes or accuracies.

ESTIMATING PLAN

The document from which all these terminology issues and dilemmas dissipate is the estimating plan, which provides the exact definition of the way each and every component of the project should be estimated.

The first choice is to what extent the estimate should be supported by a dedicated effort of studies and engineering, as opposed to using existing internal references. Let us examine some typical examples of estimating plans, their purposes and main features. In the following paragraphs, two different estimating plans are analyzed, one for a very rough estimate and the other for a binding Lump-Sum Turn-Key (LSTK) proposal. Then some examples of intermediate solutions will be given, by means of "parametrical estimates".

TWENTY-FOUR-HOUR ESTIMATE

It is a kind of support provided by experienced estimators as a very rough estimate when approaching a new project or initiative and the company senior management needs "an idea about what size it could be". The only way to do this is by referring to previous similar projects, using qualitative criteria and quantitative parameters to produce the estimate. The key points of this procedure are as follows:

- Identification of the most adequate reference case, based on (i) similar configuration (ii) similar technologies used, (iii) completion date, as recent as possible.
- Technical analysis of the most significant differences and estimating the cost impact. Here. there is no formula that can provide urgent assistance and it is only the experience of the chief estimator as well as the senior process engineers that can provide a valuable input.
- Scale up or down, due to different plant capacity. The scaling factor is based on an equation like the following:

$$\text{Cost}_1 = \text{Cost}_0 * \left(\frac{\text{Capacity}_1}{\text{Capacity}_0} \right)^k \qquad \text{(A1.1)}$$

Where Cost_0 and Capacity_0 refer to the reference case, while Capacity_1 refers to the case under consideration and Cost_1 is the result being looked for. The exponent, k is the "scaling factor", typical for each type of project/plant. The fact that k is an exponent indicates the cost ratio is different from the capacity ratio. For instance, for a complete new process unit, with a typical k factor of 0.7 and a capacity of 80% of the reference case would cost 86% of the reference as in Equation (A1.2):

$$0.80^{0.7} = 0.86 \qquad \text{(A1.2)}$$

However, if the capacity of the new plant were larger, say 20% more, then the increase in the estimate would be less than the capacity ratio, i.e., 114% higher in cost for 120% larger capacity as in Equation (A1.3):

$$1.20^{0.7} = 1.14 \qquad \text{(A1.3)}$$

Of course, the accuracy of this method is better if the capacity range considered is low and the results are becoming less and less accurate as the capacity ratio grows higher:

- Localization effect: Many "standard" cost estimates are based on US Gulf Coast costs. In the international literature, there are localization factors that will help converting a cost into some other basis, such as Western Europe, Middle East, or Far East basis.

- Time adjustment: An escalation factor should be factored in, to take into consideration the year when the reference case was built. This is usually achieved by applying indices such as the following:
  - Marshall and Swift Process Industry Index (1926: 100)
  - Nelson – Farrar Refinery Construction Index (1946: 100)
  - Engineering News – Record Construction Index (1913/1949/1967: 100)
  - Chemical Engineering Plant Cost Index – CEPCI (1957–1959: 100)
- Currency adjustment: A similar correction should be applied for the currency if the original reference cost is available in a different currency from the one needed for the new case.

The above description refers to an adjustment methodology used in some companies. However, different names are often encountered in the literature or in the practice of some companies. Some of these are as follows:

- Six-tenths rule
- Power factor to capacity ratio

LSTK Proposal Grade Cost Estimate

This is of course the case in which the accuracy of the exercise should be taken to its extreme. In the preparation of a LSTK proposal, the estimating plan is the first task to be accomplished; all other plans and schedules will unfold afterwards, as well as the proposal budget.

Every company has its own internal rules, standard plans, and procedures, some of which are more conservative and some less. As an example, let us consider requirements for a LSTK proposal grade cost estimate in most global engineering and construction contractors:

Major Itemized Equipment
The cost estimates should come from negotiated vendors' quotations, or even pre-award commitments. The definition of "major" varies, but in general it makes reference to price level and/or delivery time.

Other Itemized Equipment
At least 80% of the cost should come from vendors' quotations. The rest from sized equipment list and internal cost database of previous projects.

Piping, Instrumentation, and Electrical Bulk Materials
Based on P&I Diagrams, electrical one-line diagram and general layouts, bulk materials are "counted" (taken off) from the drawings and so the preliminary quantities are determined. Unit prices will then be taken from previously issued orders, duly adjusted for volumes and time factor.

Earth Movement and Civil Works

Based on plot plan and grading plan, preliminary takeoff for excavation quantities and preliminary foundation sizing are done. Unit prices are taken from previous projects, duly adjusted for volumes and time factor.

Pre-design

Some pieces of bulk material will require some pre-design, such as (i) special piping loops, where stress analysis may be a key factor, or (ii) critical foundations and structures, in particular, due to dynamic loads and vibrations.

Construction

This is probably the trickiest part of the entire cost estimate. The first piece of important information is generated from the quantities to be installed and is obtained as a result of the material takeoffs for bulk materials. The second and probably the most critical element is the unit prices to be applied for obtaining the total cost estimate.

- The most straight forward way to provide unit prices is to ask for them, by requesting quotations from the potential construction contractors. Therefore, inquiries should be issued to the market, complete with the following:
    - Bills of quantities
    - Description of the works
    - Reference drawings like P&I Diagrams, layouts, etc.
    - Commercial conditions (draft subcontract)
- Next step will be the evaluation of the quotations received and if possible, the relevant alignment through clarification meetings with the prospective subcontractors.

The process becomes much more efficient and reliable if it is accompanied by an alternate internal cost estimate, which can be used as a baseline to compare quotations. It further makes easier the task of coming up with a definitive price to be used in the cost estimating process:

- A very typical method for internal cost estimate is the standard man-hour method, whereby the construction cost is defined as in the following equation:

$$\text{Cost} = M \times k \times C \tag{A1.4}$$

    where:
    M = standard man-hours
    k = productivity factor
    C = man-hour all-in cost.

- Each item of construction/erection can be evaluated in terms of man-hours required to accomplish it in standard conditions.

- "Standard conditions" are defined by each company, it could be U.S. Gulf Coast, or Western Europe or similar.
- The "productivity factor" measures the efficiency of construction work, as compared with the standard conditions, i.e., the ratio between the actual number of man-hours required for a single piece of construction work, and the theoretical (standard) one.
- Therefore, M×k represents the number of actual man-hours required in the actual location.
- "Manhour all-in cost" includes all the direct and indirect costs, equipment, manpower, utilities, and management.
- Standard man-hours, productivity factors and all-in cost are important pieces of company know-how.
- If all these activities are put on a timeline, it can be easily seen how the actual material takeoff can often become available only few weeks prior to the bid closing date. Apparently, this makes it impossible to follow the above procedure, as there is no time to enquire the market. For this reason, "dummy" quantities which are some kind of rough estimates from previous jobs are often used. These are obviously not as reliable as needed with regard to the quantities, but at least can be used for receiving unit price quotations from potential subcontractors for the sake of obtaining cost information. In this way, enquiries can be issued as early as necessary after starting the proposal work. Then, once the quotations are available, it will be easy to adapt them to the actual quantities.

Temporary Construction Facilities

The so-called TCF (Temporary Construction Facilities) include temporary offices, warehousing and storage areas, camp to lodge the company personnel (if required), etc., for which a more detailed listing is given in Chapter 12 – Continuation to the Project at Field. All these facilities require to be sized based on a mobilization plan, material delivery plan, and a schedule of construction management and supervision, as well as owner's requirements (e.g., space for owner's personnel). Once the preliminary sizing is done, an inquiry can be issued for the required facilities and the corresponding prices can be obtained from the market. The TCF cost should also include an evaluation of the running costs for the whole period of construction (utilities, catering, etc.)

Home Office Services

Here is another area where commercial estimators and technical departments often disagree:

- The number of engineering man-hours necessary to execute a job must be independently estimated by both and then somebody with authority to decide, like the commercial director, has to make a decision, often mediating. It is obvious that people who will have to execute the job would prefer a comfortable man-hour budget, but sometimes it is necessary to challenge the structure to be aggressive for enhancing competitiveness. Therefore, the main issue around the man-hour estimate is not entirely technical but rather "diplomatic".

- Non-engineering man-hours: These type of man-hours, such as the ones for project management, project control, supporting services, etc., can be estimated on the basis of the time schedule, by building an assignment chart on which each position is associated with a bar indicating the duration of assignment. This method is quite objective and usually gives good results. Part-time assignments can also be represented by fractions, for example, ½ planners for the whole duration of 36 months would correspond to 18 man-months.

Once the man-hour quantity has been estimated, the next step is to evaluate the cost. This is usually done based on standard hourly cost that represents the actual cost of a man-hour of work in a given department, including the following:

- Average cost of salaries
- Social burden incidence
- General departmental costs or the incidence of indirect costs for each technical department (e.g. departments head, secretarial services, non-productive man-hours, etc.)
- General overhead costs, or the incidence of the costs outside of the technical departments (general management, commercial costs, administration and finance, human resources and other staff functions, leases, interests, insurances, and so on). This is usually expressed as a percentage applicable to man-hour costs

Field Services

Field services consist of construction management and construction supervision. They are also estimated on the basis of a time schedule and an assignment chart, which will easily provide the number of man-months to be spent at site. The monthly cost must be determined considering the international component (salaries, expatriating incentives, travel costs, rotation costs) as well as the local ones (living and lodging, visas).

Other Costs

There are a number of other costs that need to be factored into the overall project cost estimate, the main ones being the following:

- Bonds, bank guarantees
- Insurances
- Commercial costs
- Local taxes

Total "Above the Thick Line"

This is an expression used often to identify a sub-total of direct costs, basically the ones described above. Above this ideal line are the direct costs, which can be estimated on technical basis; below the line are other component of cost that are estimated in a different way, being often the result of commercial decisions. Such is the case, for instance, of contingencies.

Re-visiting the above points one by one, it becomes clear that there will be a certain amount of proposal engineering work required to support the cost estimate. For example, if the estimating plan mandates the takeoff of piping materials from the P&I Diagrams, then these must be made available and somebody must prepare them. If the plan requires 80% of equipment costs from vendor quotes, then some work must be done to prepare requisitions, issue inquiries to the market, evaluate the answers, and so on. All this is known as "proposal engineering" and it is a direct consequence of the estimating plan. This element is crucial for the proposal planning, but it also impacts the overall commercial budget for the year. It should be considered that a large proposal can imply a substantial amount of engineering work and so one or two of these big jobs could blow out the entire proposal budget for the year. A large proposal for a complex LSTK project could easily require an effort of 40,000–50,000 man-hours.

Parametrical Estimates

The variety of cost estimating approaches in between the "24-hour estimate" and the "LSTK proposal" is very wide and so, describing it in full details would be quite a difficult task. A criterion often used is the parametrization, whereby a number of items could be estimated on a percentage basis, referred to another element of the cost estimate. The most popular examples can be described as rules of thumb (heuristics). In the following examples, percentage ranges are not always provided, as these are typical of each type of project. When provided, it must be understood that the figures used are for reference only.

- Bulk materials can be estimated as a percentage of the cost of equipment.
- Cost for home office services is a percentage of the total EPC value, usually in a 10%–15% range.
- Engineering cost can be estimated using a standard number of man-hours per piece of equipment; needless to say, the figure varies from company to company, but it can be stated to be between 600 and 1,000 man-hours.
- The cost of the various engineering disciplines is often a typical percentage of the overall engineering cost, e.g., ratio of process engineering to total engineering, piping to total engineering, etc., can all be selected, if needed, by referring to historical data.
- Man-hours for project management and indirect home office services are a percentage of engineering man-hours, typically 10%–20%.
- Cost for site services is often evaluated as 10%–15% of the direct construction cost.

Many other examples could be given, but it is important to keep in mind that these are only indicative ranges and the parametrical criteria cannot be used to build up an accurate cost estimate. Main focus is on building budget estimates without spending too many estimating man-hours and above all, on providing cross-checks for costs determined by using more detailed methods. If a cost falls grossly outside the range, one should wonder why and no stone should be left unturned until an explanation is found with the figures reconciled. This sort of cross check is often referred to as a "sanity check".

COST ELEMENTS "BELOW THE THICK LINE"

As anticipated, some items remain to be still estimated, once the direct costs are determined on technical basis. This is done ideally below the thick line and often consists of a series of percentages.

CONTINGENCIES

In risk management, a contingency is an amount of money allocated to cover for unforeseen events that may occur, causing additional costs that would otherwise immediately turn the financial result into the negative. Many distinguish between non-quantifiable costs the occurrence of which is anticipated and those which are rather unpredictable:

- The first category is mostly represented by the so-called Technical Development Allowance (TDA), a percentage of growth of the cost of equipment and material, which will inevitably occur, no matter how accurate the estimate is. This concept is verified statistically and it can be proved that a certain amount of cost increase will occur in any case; of course, the level of TDA percentage depends on the method used for the estimate of that particular cost. For instance, the level of TDA for an item estimated on the basis of a purchase order issued will be lower than it would for an estimate based on statistical data, as more development work can be expected in the latter case. Usually TDA are built into the base material costs and are left above the "thick line", because they are considered as an integral part of the equipment and material cost.
- The second category includes the real contingencies, to cope with cost increase due to unforeseeable events. Some examples are as follows:
  - Estimate errors or inaccuracies
  - Inaccurate understanding of contractual obligations
  - Variations of scope not compensated for by the owner by means of change orders
  - Owner's negative attitudes
  - Unexpected market situations due to international crisis or other factors
  - Political turmoil
  - Site conditions different from original assumptions
  - Construction subcontractors' lower-than-expected performances

Each of these conditions must be identified in the bidding phase by way of an accurate risk analysis and covered by a contingency amount which is to be used only if and to the extent that the risks are not mitigated in other ways.

As a general rule, the more accurately a proposal is prepared, the less contingencies will it require. The use of contingencies is necessary, but it must always be kept in mind that they can kill a proposal, i.e., if contingencies increase the price dramatically, the prospect will consequently be lost. To give an example, in a LSTK proposal there is no point in spending a lot of man-hours to accurately estimate equipment cost when construction costs are estimated without inquiring potential subcontractors. In this case, at the final price review meeting, the commercial director will have no choice but to add extra contingencies that are very likely to kill the proposal and nullify the efforts made.

Another point to be kept in mind is that contingencies are costs and as such they must be properly dealt with. A good project manager will try to save on project costs and likewise strive to avoid occurrence of the negative events that would force him or her to "spend" the contingencies. In case the risk does not materialize and the negative event does not occur, the relevant amounts of unused contingencies can become a saving and increase the project margin.

RISK AND PROFIT

This is a factor that the senior management establishes by defining the amount of money that the company desires to earn as a profit from the project. It is sometimes called "initial risk and profit", since the actual value varies during the project life and the final amount will only be known at project closure. The fact that the same factor combines risk and profit is an indication of the entrepreneurial dimension of the company. The "risk" does not correspond to the ones covered by contingencies but are those that may arise from contractual liabilities for the execution, such as liquidated damages.

ESCALATION

Costs estimated belong to the time the estimate is carried out, in other words, they have a certain limited validity period, often insufficient to cover the entire project execution period. For example, home office services are estimated using the standard cost for the current year, but the project is going to last – say – 3 years, therefore, it is necessary to make provisions for the second and third year, against the inherent growth of costs. Likewise, when costs of equipment items are estimated on the basis of vendors' quotations, these have a validity period; if any purchase order is issued after the expiry of such validity, then it is normally expected that the vendor's price will likely increase.

PROCUREMENT GAMBLE

When the contract is awarded, the project manager receives a budget from the company senior management. During the execution, the actual expenditure may increase due to unexpected reasons, but it may as well be reduced if the procurement campaign is carried out in an effective way, e.g., by obtaining discounts on the prices previously quoted. Such possibility can be contemplated since the beginning and the potential discount can be evaluated right at the proposal phase and factored into the cost estimate as an anticipated saving.

NEGOTIATION MARGIN

Depending on the award process devised by the owner, it may be necessary to negotiate the price and give discounts in order to secure the contract. Such discount, in principle should not erode the level of profit expected and this is why an amount of money is set aside in the original budget to cope with such circumstances. This is also a limiting value for the negotiation team, as to how much money they are allowed to "leave on the table".

VALUE ENGINEERING

At any time during a project, a value engineering exercise can be carried out. This is a structured process to seek cost saving in a proactive and co-operative way by thinking in "out of the box" style. The exercise often involves both parties to the contract, i.e., the owner and the contractor and is aimed at identifying which parts of the project could be modified in order to save money, while respecting the general objectives set forth by the owner for the project. Such efforts may include the following:

- Introducing smarter technical solutions
- Expanding the project vendor list to include more competitive vendors
- Downgrading general specifications and waiving stringent conditions imposed by codes and standards. A typical example is the use of manufacturer's standards as opposed to more expensive American Petroleum Institute (API) standards for pumps. The real need for API standard for all pumps should be discussed
- Removing some non-crucial parts of the project (de-scoping), using across the fence solutions. For example, buying technical gases (nitrogen, oxygen, etc.) instead of building a gas production plant as a part of the project
- Changing redundancy and sparing criteria
- Changing spare parts philosophy

Of course, it is essential that this be done without affecting safety, reliability, operability, and maintainability. The technique used for value engineering is the so-called brainstorming method. This consists of putting together a team of people with the required competences, like process engineers, project manager, specialty discipline engineers, operation people, and cost estimators.

In the first phase, everybody will be proposing solutions off the top of his/her mind, in a creative and sometimes, even disorderly manner. A facilitator gathers the proposed ideas, listing them up, and encouraging the attendees to propose more. The added value of this phase is the richness of the contributing ideas, which should be proposed without limitation. During this phase, first discussions can take place in the group and some ideas may be discarded right away, while some are retained for further analysis.

At the end of the first phase, the facilitator has a list of ideas and the second phase of screening starts, whereby each idea is analyzed with pros and cons, with the dialectical contribution of the group; only the real feasible solutions pass this screening.

Then, the third phase starts. This serves to provide a rough cost estimate and evaluate if the amount of the saving anticipated is worth implementing the solution. The final value engineering report must be approved by the owner before it may be transformed into changes to the contract scope. Negative change orders should then ensue, entailing price reduction.

This procedure is sometimes used when the project costs are too high and the feasibility is at stake. In these cases, value engineering is often a last resort for the owner to "save the life of the project".

When value engineering is performed during a contract execution, it is customary to introduce some sharing scheme for the savings obtained. This helps incentivizing the contractor to perform an effective exercise, increasing chances to obtain a really interesting reduction of costs for the benefit of both parties.

INTERNATIONAL PROFESSIONAL ORGANIZATIONS

International organizations and associations differentiate a cost estimator from a cost engineer. The difference is subtle; basically, a cost engineer's field of activities stretches out further, toward cost management and cost control.

Main international associations are of U.S. origin; among them the following can be cited:

- American Society of Professional Estimators: www.aspenational.org/
- AACE International – Association for the Advancement of Cost Engineering: http://web.aacei.org/. It publishes Cost Engineering, a monthly journal
- The Association of Cost Engineers (UK based): www.acoste.org.uk

These associations generally have programs of conferences, training sessions, best practice enhancement, and certification and accreditation of chartered cost estimators and cost engineers.

EXAMPLE OF A COST ESTIMATE

The following figures provide an example of cost estimate. The scenario considered is a LSTK proposal for a refinery plant, by a hypothetical contractor to a client in a fictitious country. Tables A1.2–A1.5 contain details of the cost estimate for four different cost categories as follows:

- Services
- Materials
- Construction
- Other costs

TABLE A1.2
Base Cost Details – Services

Fester Consulting Co.

Proposal for Refinery Unit in Magnolia – Cost Estimating Table

Base Cost Detail – Level 1 – Services

		Thousand US Dollars	
	Home Office Services		
	Engineering	4,500.00	
	Project management	1,500.00	
	Procurement	1,050.00	
	HO construction	150.00	
E1	*Total home office*		7,200.00
	Site Services		
	Site management	450.00	
	Supervision	2,700.00	
	Commissioning	300.00	
	Start up (assistance)	180.00	
E2	*Total site services*		3,630.00
E	**Total services**			10,830.00

TABLE A1.3
Base Cost Details – Materials

Fester Consulting Co.

Proposal for Refinery Unit in Magnolia – Cost Estimating Table

Base Cost Detail – Level 1 – Materials

		Thousand US Dollars	
	Equipment		
	Pressure vessels	15,000.00	
	Machinery	7,500.00	
	Electrical	1,050.00	
	Instrumentation	900.00	
	Packages	2,550.00	
	HVAC	450.00	
P1	*Total equipment*		27,450.00
	Bulk Materials		
	Piping	9,000.00	
	Instrumentation	4,000.00	
	Electrical	4,000.00	
P2	*Total bulk materials*		17,000.00
P3	Transportation		3,000.00
P	**Total materials**			47,450.00

TABLE A1.4
Base Cost Details – Construction

Fester Consulting Co.

Proposal for Refinery Unit in Magnolia – Cost Estimating Table

Base Cost Detail – Level 1 – Construction

		Thousand US Dollars		
	Works			
	Earth moving	1,500.00		
	Civil works	3,000.00		
	Buildings	1,500.00		
	Mechanical erection	9,000.00		
	Electrical and instrumentation installation	3,000.00		
C1	*Total works*		18,000.00	
C2	Temporary Construction Facilities		450.00	
C	**Total construction**			**18,450.00**

TABLE A1.5
Base Cost Details – Other Costs

Fester Consulting Co.

Proposal for Refinery Unit in Magnolia – Cost Estimating Table

Base Cost Detail – Level 1 – Other Costs

	Thousand US Dollars		
Various Costs			
Insurances	200.00		
Bond	90.00		
Commercial	450.00		
Miscellanea	500.00		
Total various costs		1,240.00	
Contingencies			
Technical Development Allowances	300.00		
Construction quantities	360.00		
Estimate contingencies	1,800.00		
Total contingencies		2,460.00	
Grand total other costs			**3,700.00**

In a real case, the input to the tables is provided by other tables or spreadsheets, containing more detailed information. So, for example, the value 15,000.00 for the "pressure vessels" category in Table A1.3 is supposed to come from another table

TABLE A1.6
Cost Summary and Pricing

Fester Consulting Co.

Proposal for Refinery Unit in Magnolia – Cost Estimating Table

Summary and Pricing

		Thousand US Dollars	
Services	10,830.00		
Materials	47,450.00		
Construction	18,450.00		
Total base costs		76,730.00	
Various costs	1,240.00		
Contingencies	2,460.00		
Total other costs		3,700.00	
Grand total costs		80,430.00	
Risk and profit (5%)		4,021.50	
Negotiation margin (2%)		1,689.03	
Posted price			86,140.53
Say			**87,000.00**

where pressure vessels are listed one by one and individual prices are shown, along with their origin (e.g., firm quotation, internal estimate).

Table A1.6 shows the costs summary and the pricing elements.

Needless to say, the figures in the various tables are totally fictitious and are intended for the purpose of exemplification. The criteria for contingencies, TDAs, risk and profit usually vary from a contractor to another, but the basic structure of the estimate will not be much different from the one shown.

Appendix 2
Feasibility Studies

A feasibility study is a systematic evaluation and powerful analysis carried out for a project or a business idea in order to get assured about its viability. It is a major reference in making the decision for either proceeding further with or quitting elaboration of a project or a business idea. As the objective of this book is to describe process facility projects, the focus will be on such type of projects here.

A feasibility study is always compiled in form of an exhaustive report, named as "Feasibility Report," covering all the surveys, collected data, integrated expertise, reviews, evaluations, assumptions, results, conclusions, and recommendations. The focal intent is to helping investors and prospective financing organizations to have an ample opinion regarding the probability of success of the proposed project.

Feasibility studies are initial assessments, involving future forecasts over a period, normally of about 15–25 years after startup of the production, based mainly on the following:

- Project information, strategies, and requirements stipulated by the investor for the project
- Historical developments in supply and demand of raw material and product markets
- Price and cost scenarios for raw materials and products (e.g., low, medium, and high or low case, base case, and high case)
- Secondary influences on price forecasts (e.g., price forecasts for other regions and relationship to related and derivative products)
- Economic growth projections
- Opportunity and risk identification
- Estimates on start and completion time periods of the capital investment
- Capital investment and operating cost estimates based on preliminary information provided from third parties (e.g., licensors, contractors, vendors)
- Climatic and soil conditions of the land
- Current assessment and future prediction for the project location (e.g., availability of transportation, energy and labor; social and political conditions)
- Financial assumptions (e.g., interest rate, debt: equity ratios, taxation)

Feasibility studies are carried out as part of activities performed along development of investment ideas, during which the information available about the investment is apparently limited. On the other hand, the level of precision and the credibility of the study depends on availability of sufficient information as well as the objectivity of the approach in making forecasts and predictions. In case there are notable missing prerequisite project information, a Pre-Feasibility Study can be conducted to help sorting out alternate scenarios before proceeding with a full study.

When a feasibility study is properly accomplished, such a preliminary assessment is yet the only and a quite valuable way to lead to judgment for the likelihood of the project's success. Hence, it is unthinkable to avoid performing a detailed feasibility study, especially when the project is large or complex, for which the corresponding report may be over some hundreds of pages. A "bankable" feasibility study is the one that possesses the contents and the level of details acceptable to financing organizations for making detailed reviews in order to decide whether to finance the project or not. Financing organizations apparently expect that the feasibility report is prepared by a company that has proven expertise in preparing such work on that particular subject of the investment concerned.

CONTENTS OF A FEASIBILITY REPORT

A feasibility study for a process facility targets assessing markets, alternative business scenarios, sensitivities, assets to be financed, profitability and funding potentials in order to ensure ending up with results and conclusions that can guide the investor as best as possible.

In cases when the process technology has not yet been selected, the study may include highlights about alternative technologies in order to provide some initial support to the investors in that respect as well. After all, it is up to the investor to go for or quit further developing the nominated project.

A well-prepared feasibility report mostly covers the following areas of concern and interest.

BUSINESS OVERVIEW

This section covers historical background of the project and an industry overview outlining types/specifications of raw materials, products and processes currently in use in the industry.

MARKET DYNAMICS

It is about supply, demand and trade outlook based on the trends in the selected industry. It highlights current global demand, supply, expected growth of demand by regions, sales projections, forecast for supply including production through new projects on the way, regional supply-demand balance, net trade and potential buyers.

PROFITABILITY AND PRICING

This section mentions about pricing influences, price history, historical profitability analysis, profitability projections, price forecasts, and risks to the price forecasts.

TECHNOLOGY REVIEWS

Overview of traditional processes, outline of different technologies offered by licensors including recent developments in the industry.

Capacity Analysis and Suggestions

Feasibility reports may also verify any pre-selected production capacities and may make suggestions for optimums and world-class scales.

Project Costs

It covers estimates for fixed capital investment (FCI) costs plus working capital (WC), collectively termed as capital expenditures (CAPEX), as well as manufacturing costs also referred to as operating expenditures (OPEX).

Regarding CAPEX, all cost items including

- Engineering, procurement, and construction contracting services
- License fees
- Purchased equipment and bulk materials including system packages and commissioning and start up spares
- Construction management, construction labor, costs of construction equipment use, consumables, and overheads for the following:
  - Site development, buildings, structures, underground installations
  - Construction of mechanical, piping instrumentation-controls and electrical systems
- Project management
- All other home office and site expenses
- Owner's all expenditures for the project
- Training, commissioning, and startup costs
- Contingencies for both the process units (inside battery limits – ISBL) and utilities/offsites (outside battery limits – OSBL)

are to be considered. In addition to the above outlined FCI costs, an operating capital (WC) cost is also estimated and used in the study.

For the OPEX,

- Direct (variable) costs including raw materials, utilities, operating supplies, operating labor
- All other direct costs like maintenance and royalties
- All fixed costs including depreciation, plant overhead, insurance and local taxes
- General expenses, such as costs for administration including general overhead, R&D, distribution and selling

are to be estimated.

These expenditures are then compared with information, if available, on existing plants.

Regarding the cost estimating, it should be noted that above costs may as well be listed or categorized in some different ways; however, care should be taken not to miss or double-count any single item.

Economic/Financial Evaluation

A financial model is developed by drawing results from the collected data and the estimates covered in previous sections of the report. The basis of the evaluation is established by making experience-based assumptions on the following:

- Time it takes for engineering and construction as well as well as the timing of the startup
- Pricing scenario (often the medium price levels are used as the base case)
- Capacity usage (normally between 90% and 100%)
- Technology configuration (a well-proven technology still in use is normally considered if technology has not yet been selected by the investor)
- Raw material and product prices, based also on projections
- Days and hours of operation per year (usually 330 days corresponding approximately to 8,000 hours per year on three shifts per day basis for most continuous processes)

Furthermore, the basic elements of the model are to include the following:

- Distribution of the CAPEX over the years (usually the first 2 or 3 years)
- Selecting an operating life of the plant (mostly 15–25 years after startup)
- Salvage (residual) value of the facility at closing down the plant (sometimes neglected, considering it pays off the costs for demolition)
- Selection of a depreciation method and period (usually, straight line over the first 10–15 years after startup for the sake of simplicity)
- Assuming reduced production capacities within the first 1 or 2 years of operation
- Corporate taxation level and exemption periods, if any, plus custom duties unless exempted
- Applying discounted cash flow analysis by using selected discount rates (normally between 10% and 18%)

By using the inputs of the model, a cumulative discounted cash flow is generated and financial/profitability parameters, like Payback Period (PBP), Net Present Value (NPV), and Internal Rate of Return (IRR) are calculated as the major decision parameters for the investors. Furthermore, some of the principal estimates and assumptions in the economic performance model of the project are verified by making up a number of sensitivity cases for which impacts on profitability are calculated for a range of positive and negative variations of these values (capital expenditure, debt: equity ratio, raw material pricing, product pricing, etc.).

Competitive Analysis

This section discusses delivered cost competitiveness of the products of the plant that changes from one region to another, by comparing with the conditions of other producers. This is performed in form of brainstorming to verify major cost contributors, like raw materials and labor. Capital costs also vary due to location factor as well as freight costs for both the raw materials and products and may significantly affect the competitiveness.

OTHER ITEMS

Other items are occasionally included in feasibility reports in order to highlight rather indirect matters but can still provide some strong help to the decision makers. These are mainly the effects of the investment on the region as well as effects of the region on the investment, such as economic growth, employment, social effects, political standing, and future projections and environmental considerations.

DISCUSSIONS AND CONCLUSIONS

This section discusses important findings of the entire study that deserve particular attention along further assessments for the final decision. A number of conclusions are also included for explicitly indisputable matters of concern. In most cases, a SWOT (Strengths, Weaknesses, Opportunities, and Threats) analysis is also made.

CASH FLOW

A Cash Flow Diagram (CFD) shows transactions that involve investment costs born and revenues obtained as a function of time along stages of a project from its initiation up to closing down the facility at the end of its commercially useful life. During investment phase, cash flows are negative due to spending monies for the following:

- FCI costs
- WC to be allocated just before starting the operation

Cash flows along plant operation after startup can be expressed as follows:

$$\text{Cash flow (after tax)} = \text{Net profit (after tax)} + \text{Depreciation} \qquad (A2.1)$$

$$\text{Net profit (after tax)} = \text{Revenue} - \text{Operating expenses} - \text{Income tax} \quad (A2.2)$$

$$\text{Cash flow (after tax)} = \text{Revenue} - \text{Operating expenses} - \text{Income tax} + \text{Depreciation} \qquad (A2.3)$$

Regarding industrial facilities, cumulative CFDs (shown in Figure A2.1) are commonly used. These diagrams start with negative cash flows cumulated in the beginning of a project due to monies spent for CAPEX and WC until the plant becomes operable. After starting the operation, revenues are expected to start generating normally positive discrete cash flows along the rest of the plant life. Hence, the accumulated negative cash flow position begins to get diminished year by year and is expected to pass to positive values as the plant is kept operated before it is finally

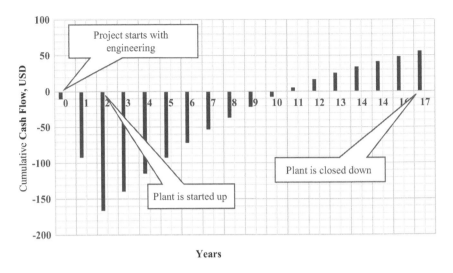

FIGURE A2.1 A typical cumulative CFD for an industrial investment project.

closed down. Although facilities are normally operated for many decades with maintenance and replacement of plant components as long as possible, feasibility reports select a plant life of around 15–25 years for the purpose of simplifying the study in this respect.

In feasibility studies, the cash flow analysis is almost always made on discounted basis. The difference between the discounted and non-discounted criteria is that the former considers the time value of money and so, each yearly cash flow is discounted back to time zero (i.e., the time of project start). For such an approach, a discounting rate which is actually an interest rate that represents the time value of money for the investor is used. Time value of money means minimum accepted profit level by an investing company for the investment against various other options of investing the same amount of money.

PROFITABILITY CRITERIA

There are basically three criteria widely used in profitability analysis. One of them is a time criterion, named as Payback Period (PBP). The other is a cash criterion, known as Net Present Value (NPV) and the last one is an interest rate criterion referred to as Internal Rate of Return (IRR).

PAYBACK PERIOD

PBP corresponds to the "time required, after start-up, to recover the fixed capital investment with all cash flows discounted back to time zero." Cost of land and WC are not taken into account since these costs are recovered at the time when the plant is closed down.

Net Present Value

NPV is the cumulative discounted "cash position at the end of the plant life." It is the cumulative cash amount obtained at the end of production life of the facility plus the values of land, WC, and any salvage that can then be recovered.

Internal Rate of Return

IRR is the "interest rate at which the NPV becomes zero after discounting back all the cash flows using IRR as the interest rate." Accordingly, it indicates the interest rate (or the discount rate for time value of money) when no accumulated money value (as NPV) is created from the investment, other than collecting such interests at the rate of IRR during the years of plant operation.

SWOT ANALYSIS

Within feasibility reports, a SWOT analysis is often included to highlight major pros and cons of constructing and operating the prospective facility. SWOT is an acronym for "Strengths," "Weaknesses," "Opportunities," and "Threats." It is identification and structured evaluation of favorable and unfavorable factors that are either internal or external to the investing company and may affect achieving a successfully running investment. Strengths and weaknesses are regarded to be generally based on the particular business, organization or the project and so, deemed to depend on internal factors. On the other hand, opportunities and threats are considered to be generally based on the environment of business or project and depend on external factors.

Appendix 3
Contracting Strategy

WHAT IS A CONTRACTING STRATEGY?

The phrase, Contracting Strategy, seems to imply a complex concept understandable to adepts only. In reality, if there is a concept that is encountered every day in our lives, even the ones not related to any technical or professional occupation, well, this is the contracting strategy.

Let us imagine an activity at home such as restructuring a bathroom. Don't we start asking ourselves how to do it? Who can do it? In how many pieces should the work be split? If we buy the tiles and provide them to the installer, maybe we can save some money but then? What if something goes wrong with the tiles and we have to replace some of them? The installer will be pleased to re-install them, but for an additional price; the chain of responsibility will be spoiled to save some dollars. And, should we go to that showroom just around the corner and order the complete restructured bathroom "turnkey"? Of course, this will cost more than splitting the work among various specialists, the masons, the plumbers, and so on, but how many headaches will this save us? There is not one single answer, it depends, for example, on how much money we can spend, but also on how much time we can dedicate to manage the interfaces among the various suppliers. Moreover, once we decide to give a contract to the showroom, how to pay them? Should we keep some money in our pocket to make sure they will do the tidying up and fix the inevitable problems, like leakages or other?

To decide all this, don't we sit with our partner and start evaluating pros and cons, planning, comparing, in one word, strategizing?

A possible definition of contracting strategy is "a strategy to define how to meet the requirements expressed in a requisition or in a complex initiative". It shows how the project will be implemented and the characteristics of the contracts that will encompass it. It determines the success of a project because it takes into account the risks and the best way to mitigate them.

BASIS OF THE CONTRACTING STRATEGY

There are so many aspects to be kept in consideration while defining the contracting strategy for a process plant project, such as the following.

SAFETY

Every solution we ponder may lead to a different execution plan and each of them will have different implication for safety of people who will implement the plan.

Think, for example, of a crowded construction site resulting from wrong planning and subsequent interference problems among construction subcontractors being simultaneously at work on the same premises. Such interference could be avoided by different measures, such as a different subdivision of the site works into work packages, or other means to minimize simultaneity of presence at site.

TIME/COST BALANCE

There is no straightforward answer to the question "where is the priority?" If a crude oil producing installation is capable of generating profits in the amount of $500 MM per month and its cost is in the range of half a billion, then would it not be wise to try and complete it 1 month earlier even though the investment cost would go up by 10%?. The terms of the problem could be radically different for a capital-intensive project, i.e., a project in which the total investment cost is the key factor in the financial model. In this case, it would probably be wiser to select a contracting strategy that privileges fixed capital investment costs, even at the risk of stretching a bit the construction schedule.

AVAILABILITY OF RESOURCES IN NUMBER AND SKILL

For example, is there a single contractor who can take the entire job and execute it safely and successfully? The answer to this question could be the basis to decide whether to use a single contractor or split the job among several contractors. Another decision that could come from this factor is to consider bringing foreign contractors into the country where the plant will be built. Therefore, this consideration takes us straight into the next point.

LOCAL CONDITIONS AND REGULATIONS

The country where the project will be implemented may have restrictive measures by law to prevent foreign contractors from working in the country, or there may be work visa problems for foreign workers. Often, it is considered prudent to reserve a portion of the job for local contractors. The law may not impose this, but it is deemed as a good approach to smoothen up social relations in the area where the plant is to be built.

FINANCING ARRANGEMENTS

In the presence of some financial arrangements, it is customary to impose certain minimum amounts of purchases to be made in a given country. One can see how this can lead to higher costs if the funds come from an "expensive" country and this aspect needs to be carefully factored into the project budget since the time of proposal preparation.

Execution in a Joint Venture

If the contracting entity is a joint venture formed by two or more contractors, then everything should be mediated, because different contractors may have different approaches for matters like the following:

- The risk inclination
- The "typical" execution and contracting scheme
- The vendor list for equipment and materials
- The project procedures

In many cases, even the cultural approach could be different. For example, in the 1970s and 1980s, Japanese contractors had very little inclination toward international joint ventures, because it was difficult to reconcile cultures, especially at the level of the employees who form the fabric of the company.

ELEMENTS OF A CONTRACTING STRATEGY

In consideration of the points above and others not mentioned, the contracting strategy to be established by an engineering contractor should include a number of items. In the following pages, a description of the most typical ones provided for the purpose of exemplification.

Breakdown in Elements of Work (Packages)

Some considerations leading to this choice have been mentioned earlier in this chapter. Another major element may come from the risk analysis, suggesting to minimize the risk in some critical areas, despite of introducing increased costs. For instance, if it is known that the mechanical erection of the plant will be the most critical factor for whatsoever reason (e.g., scarcity or volatility of resources), then it may be decided to split the contract into two. In this case, two different subcontractors will be selected to erect two different parts of the project, with the flexibility that either of them can pick up the entire job, should the other fails to perform well or to continue. This solution would certainly cost more as economy of scale is left out, but with an invaluable advantage if one of the two subcontractors is clearly incapable to do his job, or becomes insolvent or even goes bankrupt. In this case, the transition would be smooth and major disruption to the project can be avoided.

Number and Location of Project Offices

Nowadays projects tend to become more and more complex and a multi-execution strategy is often recommendable. As a general rule, Engineering-Procurement-Construction (EPC) contractors are based far away from the location of the construction site and the owner's main office. Moreover, contractors have often several design offices in various locations around the world and often, the size of the project imposes the cooperation of more than one design office, adding complexity and increasing requirements for coordination, interface management, and integration.

In many cases, local constraints exist that push the contractor to open a local office, unless of course it already has one. Constraints may be operational – e.g., an easier coordination with the owner and with the local market for material and construction – but in many cases they can come from legal provision such as "buy local" (local content) policies.

EXTENT OF REQUEST FOR QUOTATION DEFINITION

When a Request for Quotation (RFQ) is issued, a certain technical definition is obviously included in the Tender Documents. From that point on, further development of design will be the responsibility of the contractor, who will decide to what extent to carry out detailed design in order to improve the accuracy of the cost estimate for the proposal price and minimize the budget overrun risk.

Likewise, the project owner is to decide the extent of design to be provided to bidders within the RFQ, based on similar considerations. The type of owner provided design is normally Front-End Engineering and Design (FEED) for RFQs on Lump-Sum Turn-Key (LSTK) basis. The same may range from conceptual design to basic engineering package for Engineering-Procurement-Construction management (EPCm) or only engineering services.

EARLY ENGINEERING

Another area to be considered for bringing forward the award phase is the early engineering. When an EPC contract is signed, there is a certain amount of detailed engineering work to be done by the contractor. Normally, the contractor starts the engineering activities once the contract is effective. Procurement/subcontracting activities will not start until engineering reaches a certain level of progress. This is due to the requirement to issue purchase orders based on a sufficient and correct technical definition.

Early engineering concept is not uncommon for EPC jobs and the idea behind this concept is that contractor could start its contractual engineering activities even before the contract becomes effective. This will allow bringing the start of procurement forward by a number of months and cut the overall project schedule. The benefits of this solution are important, but someone, usually the bidder, has to take the risk that for some reason, whatsoever, the contract may not become effective. In some cases, the owner takes the risk, by issuing a document with legal validity, to authorize the contractor to start and guarantee the payment for the services completed.

In some other cases, the contractor itself may make an independent decision and start engineering at risk. This is typical, for example, of contracts with high schedule risk, for which the contractor prefers to put some money at stake by starting engineering, for reducing the risk of late delivery, which could have much worse consequences. It is a typical measure of risk mitigation.

EARLY PROCUREMENT FOR LONG LEAD ITEMS

For an EPC contract, material procurement is often risky, because late delivery of materials can jeopardize the overall schedule and create delays as well as additional costs that the construction subcontractors will often not be able to recover at site.

It is often possible to identify few items, which have long delivery time (lead time) and are on the critical path; these are commonly known as Long Lead Items (LLIs). In this case, mitigation of the late delivery risk is possible if the procurement of such LLIs can start before its normal timing. Typically, the project owner may start enquiring LLIs during the EPC contract bidding phase, with the intention to award the EPC contract and the LLIs purchase order, simultaneously. There are some ways to do this:

- Owner issues the purchase order and then provides the materials to the contractor on free issue basis.
- Owner issues the purchase order and then transfers its title to the contractor who will manage the order as if it had issued the same. This requires pricing arrangements between project owner and contractor. Whatever the price and terms negotiated between the owner and the vendor, the handling responsibilities as well as the mutual liabilities will have to be transferred to the contractor.
- Owner carries out the vendor selection process and negotiates the purchase order with the vendor but leaves its finalization to the EPC contractor. This will make the transfer of liabilities to the contractor smoother and safer.

For the more common case, in which the owner does not act as above, the bidding contractor (i.e., prospective EPC contractor), tries to receive firm quotations with suitable validities from LLI vendors. This is followed by performing at least preliminary bid evaluations and carrying out initial negotiations with such vendors for firming up the prices as well as the terms and conditions for a possible future award of the EPC contract. The foregoing approach is a good practice also from cost estimation point of view since most of the LLIs are also the higher priced items which require ensuring to have at hand much more reliable cost estimates.

DEFINITION OF CONTRACT TYPE

It is already seen how the time versus cost relationships can have different solutions for different types of projects. Such considerations lead to the first critical decision to be made: Should lump-sum contracts or reimbursable ones be used? For an exhaustive discussion of this point, please refer to a dedicated section in this chapter. However, the fundamental difference between the reimbursable contract family (in numerous subtypes) and the lump-sum contract family (in quite a few subtypes, too) is in the allocation of the time and monetary risk. In the lump-sum contracts, the contractor undertakes all or most of the economic risk, as the price is fixed, but the costs will only become known at the end of the project. On the other hand, in a reimbursable type of contract, the owner will assume most of the budget risk, because it will reimburse contractor of the costs actually incurred. Regarding the time, lump-sum contracts motivates contractors better to complete the works as soon as possible to allow receipt of payments sooner. However, getting ready for a lump-sum contract takes longer time, since more detailed definition of the works is needed for this contract type. Of course, in both families of contracts there are many different variations on the theme, the main purpose of which is to mitigate the risks for both Parties.

USE OF OPEN PURCHASE ORDERS VERSUS POSITIVE SCOPE DEFINITION

Open orders are those, for which the actual scope of work is not initially defined, but the specifications and unit prices of the various components of the contract are agreed. They are also known as project agreements, whereby a price list is defined for items such as piping fittings, or IT materials and services. The agreement is negotiated once and for all at the onset of the project, in parallel to the definition of the actual quantity in the detailed engineering phase. This allows saving time if the purchase is on the critical path, as the piping materials often are. When the moment comes and the material takeoff (MTO) is ready, then an order can be issued in a matter of few days, as the selection and the negotiation were completed weeks, or months earlier. This approach fits very well with piping material items, for which MTOs – approximately three – are prepared as piping design progresses. Each consecutive MTO adds in additional materials, as new items and/or increased quantities, to be ordered immediately at completion of each MTO. Such sort of blanket ordering brings the advantage of submitting piping materials to the piping contractor in partial deliveries as early as possible to have piping prefabrication started sooner and proceeded without stopping.

GUIDELINES FOR CONTRACTS

The contracting strategy must contain some guidelines for the contract(s) – or the subcontract(s) – to be used:

Risk Allocation

There is a general principle in risk management. Risks should always be allocated to the party who can better manage them. If it is possible to do this, then the overall risk can be optimized and by being optimized, it is automatically minimized. It would be good if this principle were actually followed by all project owners; but unfortunately, this is not always the case, so often distortions in the project risk profile are introduced. A typical position would be to maximize risk allocation to the contractor, with the false hope that this would take all the problems away; but whenever a contractor takes a risk, it puts an extra amount of money in its price in order to cover for the risk. This contingency, from contractor's point of view, is just a protection and it would only be spent if and when the risk materializes, i.e., when the adverse event takes place. If, on the contrary, the risk does not materialize, this amount of money becomes extra margin, either profit or compensation for some monies lost elsewhere.

From owner's perspective, in a LSTK contract the money for contingencies is spent in any case, the moment the contract is signed, because it is a part of the overall lump-sum price. On the contrary, if the risk stays with the owner, there will be no such contingency in the lump-sum price. In this way, if the adverse event does not materialize, then the owner will save this money. From here, one can see the importance of a rational risk allocation for project owner: Keep that risk and you may save something. Transfer it to contractor and the money will be lost forever.

Type of Contract to Be Used

Based on the overall project strategy and the risk strategy, the most appropriate type of contract can be chosen. The first choice is the category of contract: lump-sum type versus reimbursable type. It will be noted in a moment that within these categories there are various subtypes, but this can be a second step choice. The detailed definition of the variety of existing reimbursable contracts is predominantly an outcome of a fine-tuning exercise, which involves analyzing the risks and mitigating them.

Roles and Responsibilities in Contract Drafting and Negotiation

The main decision is whether to make use of the internal legal/contracting department, or to employ an external law firm. Generally, external counsels are used when dealing with special and complex contractual architectures, while for simpler and/ or smaller contracts, internal functions can take care of the drafting and negotiation of the contractual text. Most companies have a contracting office, often reporting to the commercial director. This office can normally deal with the vast majority of the aspects, by consulting legal experts for issues having a typical legal character.

TYPES OF CONTRACTS

From risk allocation point of view, contracts can be divided into risk distribution and risk optimization types.

RISK DISTRIBUTION

Risk distribution means that the overall contract risks can be allocated in different ways. Portions of risk can be transferred from one party to another, but still the final result is that the overall level of risk varies to a small extent. In most of such contracts, the interests of the two parties are opposing, because things are either to one of the party's advantage or to the other. One of the consequence of this situation is that the management style of both sides will be distributive and even confrontational. Figure A3.1 provides a graphical representation of the risk distribution for four types of contracts.

The following is an outline for risk distribution contracts.

Cost Plus Fee

Here, the owner reimburses the contractor for all admissible costs, by either making direct payments to vendors and subcontractors, or by reimbursing contractor after the latter bears the costs. The difference is a pure cash-flow matter. In addition to the

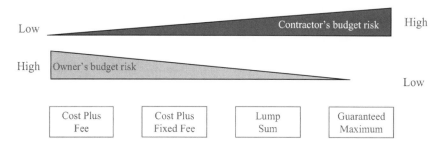

FIGURE A3.1   Typical risk distribution contracts.

direct costs, the owner will also pay the contractor a fee, calculated with a predetermined percentage over and above the costs reimbursed. In this way, if, for example, the overall project cost is $100 and the agreed fee is 7%, contractor will receive a fee of $7.00. If, on the other hand, the cost goes up to $150, the fee received will be $10.50. The example explains why project owners do not like this type of contract very much because the contractor has no incentive to reduce costs; on the contrary, it has an interest that project costs escalate, in order to be paid more.

Cost Plus Fixed Fee

The scheme is similar to the above, but there is a difference, introduced to mitigate the risk of uncontrollable budget growth. The fee is now an absolute Dollar value and no longer a percentage. Using the figures from the example above, at the beginning, the parties will agree on a fixed fee of $7.00, regardless of what the actual project cost will be. In this way, the contractor is less incentivized to raise project costs, as its fee will not increase proportionally. Moreover, if the costs grow considerably, the fee will correspond to a decreasing percentage, meaning that the effort made by contractor will be less remunerated. In fact, if the costs go up to 150%, like in the example above, the $7.00 will correspond to 4.70% instead of the original 7%.

Lump Sum (Fixed Price)

In this case, the contract price is fixed, independently from the actual costs. Therefore, the budget risk is entirely borne by contractor; of course, the latter has the possibility to include an extra element of profit and extra contingency, in order to mitigate the risk. This type of contract is often the preferred one in the Middle East and the Far East and is traditionally the best gymnasium for project managers, who have to perform with their best skills to reduce any extra costs and make profit for their company.

Guaranteed Maximum

This is a type of contract that is applied very seldom, because it is strongly unbalanced in favor of the project owner. The contract works like a cost plus fee, but there is a Guaranteed Maximum, meaning that costs in excess of the G-Max will be borne by the contractor. Therefore, the risk is potentially higher than in a lump sum, while the remuneration is predefined at the level of a cost-plus fee.

FIELD OF APPLICATION FOR RISK DISTRIBUTION CONTRACTS

Figure A3.2 summarizes some key characteristics of the reimbursable category and the lump-sum category contracts.

Figure A3.2 tells us that

- Lump-sum contracts require good definition of the scope of work and availability of deep enough details as to allow an EPC contractor to formulate its price and commit to it.
- Lump-sum contracts show also very low flexibility for changes of scope, because every addition to the original scope creates a risk of budget overrun for the contractor. This is resolvable by the introduction of the change orders.
- Contractor's incentive to perform well is higher in a lump-sum contract, because consequences of poor management will essentially affect contractor's budget, while in a purely reimbursable environment, owner will be affected.
- Reimbursable contracts require much more supervision by the owner, as the approval requirements are much higher, stretching to areas that in a lump-sum contract are the exclusive contractor's responsibility, such as detail design choices, vendor selection, negotiated cost for supplies and services, and the like.

Due to these characteristics, it is clear that a lump-sum contract can be a good option provided that

- The scope definition is good, meaning that a portion of the overall engineering design that corresponds to a well-prepared FEED, must be done (or procured) by the owner prior to entering into a lump-sum contract.
- Risks are fairly allocated in the contract to both of the Parties, based on the sound principles mentioned above.

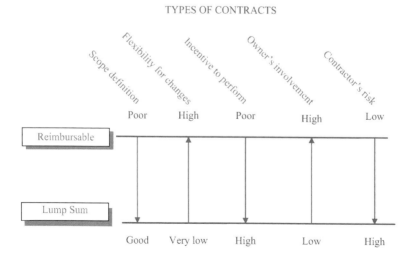

FIGURE A3.2   Field of application of risk distribution contracts.

On the other hand, there are circumstances in which the owner must award an EPC contract in a tight timeframe and there is no time to perform the required amount of engineering prior to award. This could be the case of projects with a strong time priority over budget, or projects that "must" be completed on fast-track basis for certain reasons, like pressure from the top management, urging not to lose the current market conditions, expecting changes in relevant legislation, etc. In these cases, it is probably wiser to make use of a reimbursable contract, maybe with a mitigation formula to avoid uncontrollable cost growth.

RISK OPTIMIZATION CONTRACTS

This type of contract makes use of innovative formulae, by means of which, the overall project risk is reduced, with benefits for both parties. Below are two examples: "reimbursable with incentive scheme" and "Converted LSTK (C-LSTK)".

Reimbursable with Incentive Scheme

Reimbursable with incentive scheme is the evolution of the traditional cost plus fee contract, in which the fee must be "earned" by contractor on condition for a good performance. Figure A3.3 explains the mechanism of such a contract.

Let us assume that the conditions are the following:

- Engineering services are paid with a lump-sum price.
- Materials and equipment are reimbursed by owner to contractor.
- So are construction services.

The incentive scheme is added to the above in order to discourage contractor from letting the cost for materials and construction (let us call it the Hard Cost) grow,

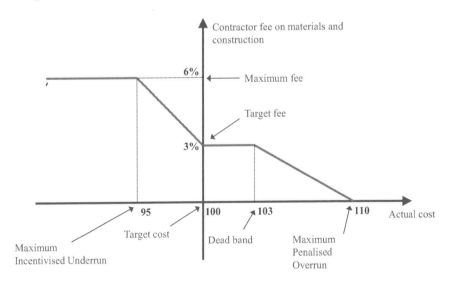

FIGURE A3.3   Mechanism of a reimbursable contract with incentives.

but also to incentivize it to make some saving – of course within acceptable limits, i.e., without compromising on the quality of the final product, as stipulated in the contract.

The incentive is set at 3%, which will be earned by contractor if the final Hard Cost will be, say 100 monetary units, as per original budget. If the final Hard Cost is higher than 100 due to contractor's poor performance, this will trigger a progressive penalization, until the contractor sees its fees zeroed if the Hard Cost reaches 110. The so-called "dead band" from 100 to 103 Hard Cost recognizes the uncertainties of the cost estimates and gives contractor a tolerance on the Hard Cost increase, before it starts being penalized by the fee reduction.

If on the other hand the contractor is able to obtain cost saving by applying its ingenuity to the detail design or by conducting a more effective procurement campaign, the fee can grow up as an incentive. Of course, there is a limit and no fee increase will be granted if the cost saving exceeds 5% (Hard Cost = 95), because otherwise this could be seen as an incentive to wild savings and quality could be in jeopardy.

The interesting point about this kind of scheme is that it generates a cooperative environment instead of confrontational, in that owner's and contractor's teams will work substantially for the same objectives and the conflictive attitude will be reduced.

Another interesting feature of the scheme in the contract formation phase is that everything can be negotiated, from the value of the base fee to the mechanisms of its reduction and increase.

Converted LSTK

This is a scheme that became popular in the Middle East around the mid-first decade in 2000s, when the international market for equipment and materials faced an almost unprecedented boom. Under those conditions, it was virtually impossible to formulate a lump-sum price in a trustworthy and reliable way, without adding massive contingency margins that would have killed every project in all other times.

Under a C-LSTK scheme, the project will start on reimbursable basis; meanwhile the contractor will build up a cost-estimating file in the most transparent way, with the owner's cooperation and control. Such exercise will lead to an Open Book Estimate (OBE) that will be sanctioned by both owner and contractor and if required, even certified by an external authority. At the time when the OBE is ready, the contract will be converted into a pure LSTK, with a lump-sum price obtained by multiplying the open book cost by a previously agreed factor that will include residual budget contingencies plus risk and profit. From that moment on, the project will be carried out under a conventional lump-sum contract. The crucial decision is "when to convert", i.e., when it is the right moment to transfer the lump-sum risk from owner (in the reimbursable phase) to contractor (into the lump-sum phase). Many consider that it should happen after 6–9 months from the effective date, but there is a point in a typical time schedule when several things happen simultaneously:

i. The engineering progress reaches 60%.
ii. About 80% of the equipment have been ordered.
iii. First piping MTO has been issued, for about 70% of the piping material.

iv. Civil construction subcontract has been issued.

v. Mechanical erection subcontract is under negotiation.

These conditions suggest that this is a good enough point to make the risk transition. Incidentally, this usually happens after 6–9 months of the effective date. However, when setting up the contractual framework, it is advisable to refer to a set of conditions rather than a date, which for various reasons could be missed.

STRUCTURE OF CONTRACTS

A complex EPC Contract is made up of a contract body plus two main annexes. The contract body is a collection of clauses dealing with the various aspects of the project, dictating rules, setting framework, rights and obligations, and so on. The annexes are used to rule on more detailed aspects, generally of technical and project management nature.

Figure A3.4 shows the various parts constituting the whole contract and is self-explanatory as far as the priority of the various parts is concerned.

Agreement (Letter of Award)

This is a very short document, essentially a letter, whereby both parties state their intention to enter into an agreement and do really make a contract. The agreement is also sometimes referred to as the letter of award and is not to be confused with the letter of intent (LOI) which is described later in this chapter.

The main elements of the contract are summarized here:

- The identification of the parties
- A synthetic representation of the scope of work

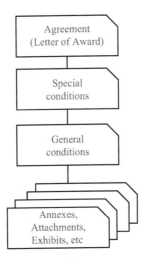

FIGURE A3.4 Typical contract structure and hierarchy.

- The contract price
- The identification and list of the main annexes with the order of precedence
- The signatures by the authorized representatives of both parties
- The date of contract signature and the date for completion agreed
- Such a document is not always present and can be replaced by few statements as the cover of the special conditions

GENERAL CONDITIONS OF CONTRACT

This is a standard document used by the project owner in all its contracts of the same nature. It touches on all the aspects on which a contract is supposed to rule and is not modifiable because it represents a sort of historical memory of the company. Any amendments to the provisions of the general conditions are contained only in the special conditions.

SPECIAL CONDITIONS OF CONTRACT

This document is set to amend the general conditions, in the form of additions, deletions, and modifications. The reasons for such exceptions can be multiple. As mentioned above, the general conditions are often an old text, which is superseded even as a standard document inside the project owner. In this case, the special conditions contain amendments to the company standards themselves. Another good reason to amend the general conditions is to introduce the result of the negotiation with the selected contractor, which of course will take priority over the standard conditions.

The special and general conditions jointly constitute the contract body. One could argue that such a mechanism of an unchangeable but superseded document (the general conditions) amended by another one (the special conditions) is useless and cumbersome and both documents could better be combined in just one. The objection is probably valid, but this is the way it is and when a contractor receives an RFQ formulated in this way, it has no other choice than to play by the rules established by the owner.

TECHNICAL ANNEX

This can be a very voluminous document, often consisting of various volumes, because it contains the technical description of the project that is the subject of the contract.

The Technical Annex contains the documents that are necessary to define the "what"s and "how"s about technical contract requirements, i.e., what and how contractor has to design and erect. Just as a matter of exemplification, this could include the following:

- Project description
- Definition of the scope of work
- Inclusions and exclusions
- Definition of battery limits of the supply

- Product specifications, including expected and guaranteed ones
- Performance testing procedures
- Specifications, data sheets, general specifications, duty specifications, instrument data sheets
- Drawings such as process flow diagrams, P&I diagrams, plot plans and layouts, electrical one-line diagram, steel structures one-line diagram, various types of assembly drawings, related existing drawings if the project is on an existing facility
- Codes and standards to be used for the design
- Information on utility consumption, usually in the form of utility summary; available utilities with corresponding excess capacities if the project is on an existing facility
- Information on effluents
- Licensor's information, in some cases a full Process Design Package by the licensor

PROJECT EXECUTION ANNEX

It contains the definition of the "how"s and "when"s about management of the project, i.e., how the owner and the contractor should interact to achieve the common goal. It should contain the following:

- Time schedule, which must always be read in conjunction with its qualifications, i.e., the conditions that must be verified in order for the contractor to be able to meet its time for completion obligations
- Progress assessment procedure with the WBS (Work Breakdown Structure)
- Payment methods
- Coordination procedure
- Invoicing procedure and other administrative procedures
- Reporting requisites and procedures, to provide guidance to the contractor in the preparation of the monthly progress report and other reports
- All other procedures applicable in the management of the contract
- Project vendor and subcontractor lists, i.e., the list of vendors and subcontractors among which the contractor is authorized to select suppliers of goods and services for the project

ORDER OF PRECEDENCE

It is necessary to establish an order of precedence among documents, to manage possible conflicts and contradictions between one document and another. The statement about order of precedence is usually contained in the special conditions and – if present – even in the agreement (letter of award). The precedence order is a relatively easy matter; when dealing with agreement, special conditions and general conditions, the order is as per this sequence.

The problems start with the other annexes and the documents that are part of those, in the form of

- Annexes
- Attachments
- Exhibits
- Appendices
- And more phantasy names...

Most claims and disputes arise out of contradictions among contract documents, which cannot be cleared with the "order of precedence" provision.

Simpler Cases

Not every contract is as complicated as that, probably just a few, when dealing with complex EPC projects. The agreement is often not present, while special and general conditions can be combined in a single document, called, for example, "the contract".

The complex architecture discussed above must be seen as a sort of checklist. What is said in 1,000 pages for a large contract can be said in 10 pages for a small one, but the principles must be there, if and to the extent they are applicable.

Letter of Intent

A special mention should be made of the LOI, a document that was mentioned few lines above. This is not to be confused with the letter of award, or agreement, which is a contract in its full capacity. The LOI is an interim document by which the owner and the contractor agree to start working even in the absence of some key conditions. These could include the following:

- Coming into force of financial arrangements such as letters of credit or project financial close, or
- Final negotiation of the detailed agreement and conclusion of the numerous documents seen above. In this way, the completion and signature of a full contract will still take time, but the project can start without further ado

In order to be a valid instrument for the intended purpose, a LOI should contain as a minimum the following provisions:

- The owner confirms the intention to enter a contract with the contractor.
- The conditions for this to happen are stated, but they are not in place at the time when the LOI is issued.
- The owner authorizes the contractor to start working, within limits stated in the LOI.
- The owner commits itself to pay the contractor for the work done within those limits. This is a key condition and many contractors would not accept the LOI unless it contains a clear commitment to this effect.

The other conditions that will constitute the contract (general and special conditions, technical and project execution annexes, etc.), may be attached or referenced, if they have been agreed. In case they were only partially agreed, then a copy of the relevant documents in their current status can be attached to the LOI. The limits of the approval depend on the constraints by the owner. Typically,

- A time limit, for instance, validity period of 3 months. In this case, the letter should state what would happen at the end of the period: Will the letter become null and void? Will the Parties have a chance to sit and discuss a renewal? Automatic solutions are never the good answer and problem solving by mutual agreement is generally the best way out.
- A maximum amount of money that owner will pay to contractor under the LOI, due to budget limitation, as the owner has not yet received all the internal authorizations and releases.
- How to deal with commitments on the market. If the works under the LOI last more than a certain period of time without transforming the letter into an official contract (say 4–6 months), then the contractor will have to stop working. Failing to do this, it would become very inefficient as it will not be able to issue purchase orders, without which it will not be in a position to incorporate vendors' data and drawings in its final design.

A purchase order may trigger little – if any at all – disbursement at the time of signature, but it certainly implies an obligation to make payments later on (for example, upon delivery of the goods to the jobsite). The level of commitment can easily exceed the amount of money that contractor is authorized to spend and so, will be paid by the owner under the LOI. To cope with this requirement, the LOI could state the commitment by the owner to honor the financial requirements of the purchase orders. Alternatively, the LOI could state that purchase orders issued by contractor will include cancellation clauses. In this case, if the LOI is revoked and the projects are terminated, the purchase order can be canceled, subject to the payment of a cancellation fee.

Appendix 4
Contractual Clauses

Scope of this appendix is not to provide an exhaustive handbook for contract writing or negotiation, nor to provide a rigorous legal background for the contract itself. This appendix is aimed at providing a general knowledge to those people who will be sooner or later involved with contract negotiation and management. Most contractual clauses have a direct relationship with project management matters, while some others have a legal-oriented content. Both must be understood by those who have or will have to manage projects and contracts. The reader is encouraged to expand the concepts presented in this appendix, by consulting specialized texts.

The following notes will provide general explanation of the most common contract clauses, as well as some hints as to the most controversial issues of discussion at contract negotiation and contract management stages. The inclusion of this appendix in a book about projects is significant for the fact that many mechanisms and dynamics in the management of a project can be explained in contractual terms. A contract sets the rules of the game and the rules for its own alterations and management. Contractual mechanisms determine the project management mechanisms.

DEFINITION OF SCOPE AND EXECUTION REQUIREMENTS

There is a clause or a series of clauses dedicated to the description of the scope of work. Of course, given the contractual structure illustrated in Appendix 3 – Contracting Strategy, most of the details for the definition of scope of work are described in technical annex of a contract and the contractual clauses make reference to it. Nevertheless, the contractual text sets the overall framework for the applicable documents, considering the order of precedence of the various parts and annexes.

One very important concept that must be carefully negotiated is about inclusions and exclusions. Obviously, in a big contract it is not physically possible to address every single inclusion and exclusion in a short clause in a section of such a contract, like special conditions. There are two approaches to deal with items that are not specified in the contractual documents. These two alternatives are described below in a very simplified and sketchy way, to make the difference more evident:

- Everything is included in the scope of work, unless it is expressly excluded
  This clause may be formulated in several different ways but in any case, it means that in case of dispute, the contractor will almost certainly lose if the disputed item is not expressly mentioned in a list of exclusions.
- Everything is excluded from the scope of work, unless it is expressly included.

In this way, the clause has a completely different meaning and the only responsibility of the contractor will be to supply whatever is actually listed among the inclusions.

Of course, neither of these extreme interpretations is acceptable. Owner will prefer the first formulation, while contractor will be happy with the second one, but in a negotiation the common sense should always prevail. Therefore, whichever approach is selected, the extreme consequences should be mitigated by making reference to acceptability criteria such as safe operation, reliability, maintainability, and the like. The reference to International Good Practices, although a bit vague, still provides at least a barrier against extreme interpretations from either side.

DEFINITION OF BOTH PARTIES' OBLIGATIONS

Both parties, the owner and the contractor, have to meet all contractual obligations and this is essential to achieve the contract's goals. The clause regarding obligation has a similar structure with the one(s) relevant to scope of work, as mentioned above. Accordingly, two extreme approaches are possible:

- In case of dispute, everything is the contractor's obligation, unless it is expressly mentioned as being the owner's.
- In case of dispute, everything is the owner's obligation, unless it is expressly mentioned as being the contractor's.

Here too, the difference is very delicate and extreme interpretations should be avoided by use of clearly expressive wording in accordance with the concept of reasonableness.

Another important point about obligations is the time factor. In many cases, stating that a party has an obligation is not enough if the timing for meeting the obligation is not specified. A typical example is owner's obligation to make the jobsite available to contractor for the construction of the plant. Such obligation will effectively not be met if the jobsite is made available too late. In this case, contractor is not in a position to maintain its obligations because of owner's failure in timely meeting the pre-requisite obligation of its own. Therefore, the suggestion is always to prepare a responsibility matrix where both Parties' obligations are listed, sometimes together with the relevant dates.

CHANGE ORDERS AND CONTRACT AMENDMENTS

The definition and the general concepts of change orders and contract amendments are discussed in Chapter 18 – Contract Administration, along with the relevant typical approval procedures. The relevant contractual clause must state that alterations to the contract provisions can only be made by the same authority that approved the original contract. The clause should then identify change orders as the only exception, as change orders can be approved by project managers. It should also contain the same concepts and outline of approval procedure as mentioned above, including details of the payment method. Change orders are often paid separately, in order not to have to revise the complex Work Breakdown Structure (WBS), forcing the readjustment of the weights of all other project activities.

PROCEDURES FOR COMPLETION, ACCEPTANCE, AND HAND OVER

This clause – or these clauses – has a crucial importance in the economy of a Lump-Sum Turn-Key (LSTK) Contract, because they deal with the most delicate phases of a project, i.e., the completion, acceptance, and hand over. Most litigations cases occur in these phases, or at least, this is when any relevant issues become critical. It is because in these timeframes most of the contractual liabilities must be discharged while entitlement to any liquidated damages are to be carefully reviewed if something has gone wrong with the time for completion or with plant performance. For contracts other than LSTK, the subject phases have similarly a great importance, but the order of magnitude of impacts to contractors of any such defaults is much less significant.

It is essential that the contract spells out rights and obligations with the outmost clarity and this must start from a clear and straightforward definition of the mechanisms for the completion, acceptance, and handing over. A summary of these typical mechanisms is provided below in a synoptic form. Figure A4.1 gives a graphical representation of the main milestones of a contract's life, on a timeline schedule.

The numbers and letters have the following meanings:

1. Effective date of the contract
A. Pre-commissioning phase—checking of the constructed plant against the contractual requirements for mechanical completion
2. Mechanical completion—relevant certificate must be issued and approved
B. Commissioning and testing period—startup and performance tests are carried out in accordance with the contract
3. Provisional acceptance—relevant certificate must be issued and approved. Usually, transfer of care custody and control occurs at this stage
C. Warranty period
4. Final acceptance—relevant certificate must be issued and approved

The sequence of events is straightforward, and it is essential that the relevant contract clause(s) represent it properly.

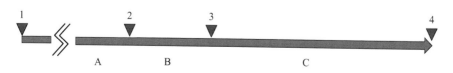

FIGURE A4.1   Synoptic representation of phases and milestones.

GUARANTEES, WARRANTIES, AND LIABILITIES

In this clause, the contractor guarantees that the plant shall meet certain levels of performance established in the contract.

PERFORMANCE GUARANTEES

These are of different nature, depending on the actual characteristics of the project. The following parameters are generally guaranteed:

- Capacity of the main product and other products. Units of measurements depend on current commercial practices for different products
- Characteristics/physical and chemical properties of product(s)
- Energy consumption, i.e., fuel gas/fuel oil consumption
- Utility consumption, either by separate guarantees on each of electric power, HP/MP/LP steam, etc., or by a compensation formula that calculates an overall consumption based on equivalence factors

SAMPLING AND MEASUREMENT

When defining performance guarantees, it is essential that the measurement methods be stated for the various parameters, including the tolerances of measurements as well as the sampling and analysis procedures. Due to the complexity and voluminous nature of the matter, guarantees and methods are often contained in a contract exhibit for purely practical reasons. The exhibit will have higher priority than any other annex, as it is deemed to be part of the contract.

REMEDIAL ACTIONS IN CASE OF NON-FULFILLMENT

Performance guarantees can be absolute or liquidable. Absolute guarantees mean that no allowance is accepted and the parameter should be obtained without any deviation. This is mainly the case for some product quality parameters that must be met in order for the product to be marketable. If the parameter is not attained at required value, then the product is useless and contractor should make good the plant with whatever measure is necessary, until the right value for the parameter is attained.

Liquidable guarantees mean that in case of non-fulfillment, the contractor may choose to pay liquidated damages in order to settle the issue. The amount of liquidated damages is calculated as a percentage of the contract price and is generally minor if compared with the actual damage that owner will suffer over the entire life of the plant. Nevertheless, in contractual terms the damage will be deemed to be finally and totally compensated by the application of that amount of liquidated damages and the owner will have no right to ask for any additional compensation.

Some guarantees have a double nature: they can be liquidable within a certain range below the guaranteed values and absolute at the low end of the range. The concept is clarified in the example shown in Figure A4.2, where the matter is explained in a graphical form, with the figures being fictitious.

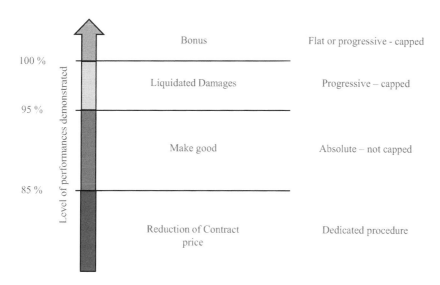

FIGURE A4.2 Remedial actions for non-fulfillment of performance guarantees.

As an example, let us consider plant capacity, which is guaranteed at say 100 [units]. If the actual demonstrated capacity is in the range between 95 and 100 [units], then the contractor will pay 1% of the contract price for each 1% of capacity below 100 [units]. If the capacity is less than 95 [units] then contractor shall have to make the plant good, with all necessary modifications, alterations and additions as are required to obtain at least 95 [units]. Once the capacity falls again into the range, liquidated damages can be applied. It is always a contractor's option to pay liquidated damages or make good the plant, at his choice.

The figure also shows two new concepts at the upper and lower end of the range. If contractor manages to obtain a capacity in excess of the guaranteed value, it will be eligible for a bonus, provided of course that this provision is spelled out in the contract. On the other side of the range, if the demonstrated capacity, notwithstanding various attempts to "make good" by contractor, still falls below 85%, then the contractor will be in contractual breach and a reduction to the contract price can be applied. Usually the extent of such contractual provisions is limited to setting forth the concept, while the quantification criteria for the price reduction are left to be agreed, possibly with an assisted procedure (e.g., mediation).

Liability Cap

Each liability should normally have a cap, which means a top limit. Liabilities of the same type (e.g., various products or various utilities) have usually an aggregated cap. Performance guarantees as a whole will have a performance aggregate cap formed by gathering relevant individual liability caps. It is to be noted that in the common contractual practice, the sum of the individual caps is always higher than the relevant aggregated cap.

Likewise, the overall performance liability cap and the delay liability cap will have an overall aggregate cap.

OVERALL LIMIT OF LIABILITY

One element that should always be present in a contract is the overall limit of liability. Most contractors will never accept a contract without a limit of liability, which should be a reasonable one, as well.

There are some liabilities which by nature, or by common practice, are not capped and they are therefore excluded from the overall limit of liability. The most typical ones are as follows:

- Make good: Contractor will have no limitation to the amount of money to be spent to make the plant good and obtain higher performances that are raised up to levels within the liquidable range.
- Patent Infringement: If contractor, during the execution of the contract, infringes the intellectual property of a third party, the owner may face the risk of a judiciary restrictive measure that would prevent commercial operation of the plant. In this case, there is no limit to the efforts and the money that contractor shall spend to reinstate the owner's right to operate the plant.
- Environmental violations: Again, due to emission problems that may be caused by contractor's faulty design, a court may order owner to shut down the plant. In this case, contractor should remedy all issues as well as any accumulated consequences without any cost limitation.

It is worth noting that the protection offered by the "limitation of liability" clauses can be lost in case of gross negligence or willful misconduct. Further details are covered later in this appendix.

SUBCONTRACTING

There are few provisions that should be stated in the subcontracting clause:

- Contractor is allowed to sublet parts of the work, provided that some conditions are met.
- Subcontractors should be chosen from a subcontractor list included in the contract as an appendix.
- Rules for addition of names in the list should be clearly spelled out. A qualification process is always required and owner reserves the right to approve the additions.
- Contractor retains responsibility on the work, even when part of it is sublet.
- Owner reserves the right to approve the selection of subcontractors as a result of contractor's subcontracting activities. This is often due to regional politics imposed, as owner has often to deal with local aspects such as manpower employment and unemployment.

DELAYS

When accepting the contract, contractor commits to execute the project within a certain timeframe. This commitment constitutes a guarantee and it may be expanded into meeting several intermediate dates, like milestones. In most cases, the guaranteed completion time corresponds to one of the followings:

- Mechanical completion, including pre-commissioning done. Punch list items are acceptable, provided they are minor and they do not prevent safe plant operation. Furthermore, there must be a commitment by the contractor to fix the punch list items before the next steps if the scope of work goes beyond mechanical completion.
- Ready for commissioning, meaning that the plant is ready to accept process fluids (and/or solids, if it is a solid-fluid or a solid processing plant).
- Provisional acceptance, i.e., plant meeting performance guarantees, demonstrated through performance tests, as specified in the contract.

The contract clause should define how to deal with delays, whether actual or anticipated. In the contracting practices of some years ago, it was customary to have only provisions dealing with actually occurred delay, i.e., plant completed after the contractual date. The remedy in this case is liquidated damages (LDs), usually spelled out as a certain percentage of contract price per week of delay, with a cap. The cap practically means that delays are liquidated up to a certain number of weeks and it is important that this limit be not reached too early. Some owners are tempted to impose a very tough schedule, hoping that this will force contractor for meeting its obligation on the time of completion. The reality is different, because if the maximum amount of liquidated damages is reached too soon, then the contractor will have no more constraint forcing it to accelerate and this leads often to ambiguities for the updated completion targets and causes further uncontrollable delays. A measure that can work as an effective deterrent is a non-linear LD schedule for delays, like in the following example:

- 0.5% per week for the first 2 weeks
- 1.0% per week for the following 2 weeks
- 1.5% per week for the next 2 weeks
- 2.0% per week for the following 2 weeks, up to reaching a cap of 10% of the contract price

In this case, delays would cost more and more for each further week and contractor will have a real interest to put in place effective recovery measures.

In modern contracts, the approach taken is more constructive and the schedule is constantly monitored by the owner. When a negative trend is spotted, owner has the right to demand corrective actions in order to change the trend, accordingly and restore chances to meet the final guaranteed date. This provision is somewhat controversial and it is often a hurdle in the negotiations, as owner would like to be entitled to demand a recovery plan at contractor's expense as soon as there are signals of possible delay. On the other side, the contractor usually claims that delay is its problem and contractual remedial actions (i.e., liquidated damages) are only due if the final delivery date is not met.

A similar way to try forcing the contractor into corrective actions against an anticipated delay during execution is to set several intermediate guaranteed dates, in connection with project milestones. In this case, the best incentive to meet milestones is to associate some payments to the achievement of a set of milestones. If a milestone is delayed, then the contractor will be automatically penalized by the impact on its cash flow.

One last interesting point is the misconception by many contractors that liquidated damages play against him and a contract would be better off without them. In reality, liquidated damages are a protection to the contractor, as well, because they constitute the sole and exclusive remedy available to owner in case of delay. If there were no such provision, then contractor would be in breach of contract in case of late delivery and this mean that owner could exercise other remedies, such as taking the contractor to court and requesting indemnification for damages, which could lead to a much more onerous bill.

FORCE MAJEURE

Force majeure is commonly defined as an event that (i) could not have been predicted, (ii) is outside of the control of the parties, and (iii) consequences could not have been mitigated.

If these conditions are satisfied, then force majeure may become a justification for not meeting certain obligations on time.

The contract clause normally contains the following:

- Definition of force majeure, based on the above concepts
- Exemplification of the same, by means of a non-exhaustive list of occurrences that can be considered as force majeure
- Definition of the consequences of a force majeure
- Right to suspend and terminate the contract if the event of force majeure lasts more than a certain number of months

In terms of consequences, the prevailing criterion is that the affected party is excused for time impact but has to bear the costs of the event. So, for example, let us suppose that a ship carrying an important piece of equipment to the jobsite sinks during a strong storm. In this case, the equipment needs to be reordered, causing a delay of the final completion date, which will be excused without any penalization to the contractor. The replacement cost for the equipment is, in any case (i.e., whether the incident is accepted as a force majeure or not), for contractor's account. The fact that insurance will pay for the loss should not lead to interpreting differently: The cost responsibility remains with the contractor; insurance is just a mitigation and does not often pay the full cost due to limits, exceptions, and deductibles of the insurance policy.

INSURANCES

The project insurance scheme is contained in a dedicated clause, typically divided into two sections: Insurances to be provided by owner and the ones to be obtained by contractor. The decision of who provides what is generally not controversial, as both Parties are anyway co-insured in all such policies through cross liability and cross-coverage clauses and therefore they have the same rights and enjoy the same protections. Moreover, in case contractor has to take out the policies, the relevant premium will then be included in the lump sum price and so there is no extra economic burden to any of the parties.

Let us analyze the main typical insurance policies provided in industrial facility projects.

CONSTRUCTION ALL RISK/ERECTION ALL RISK

This policy, commonly called CAR/EAR, covers damages to or loss of a plant that is the subject of a contract, at any stage of the construction and erection, but only up to mechanical completion.

Every single piece or component of the plant is insured, whether it is incorporated in the works, or still at manufacturer's workshop, or in transit by any means (road, train, sea vessel, airplane). Owner, contractor, vendors, subcontractors, inspectors, and any other party involved are all co-insured, i.e., they are all beneficiary in case they suffer any damage or loss of material, fabrication, supply and works either under progress or already completed. In particular, it is worth mentioning that this policy also covers owner's existing properties, i.e., those properties that are not the scope of the contract and are not included in the project, but may suffer a damage as a result of an incident on the plant under construction.

CAR/EAR is the largest insurance policy in a project, because it usually covers the entire value of the project, plus some allowances. It is already mentioned above that the decision of who should take it out is generally not controversial. Therefore, unless one of the parties has an interest in taking out the policy with a specific insurance company, the decision is taken on opportunity basis. As a rule, large contractors that execute similar projects worldwide have access to better conditions, because of the volume of business they bring to insurers. In this case, it is advisable that contractor take out the policy, as in the end the owner will take benefit of the reduced price. For similar reasons, if the owner is carrying out a major megaproject comprising several work packages, then a single insurance company would better be involved, to achieve economy of scale. In this case, CAR/EAR for the project may even be an extension of an existing policy.

THIRD-PARTY LIABILITY

This is generally a section of the CAR/EAR policy covering damages to third parties originated from one of the co-insured entities. It is noteworthy to analyze the difference between owner's existing properties and third party's properties, which in some case may become subtle. Let us imagine that the project consists of a new unit inside an existing refinery owned by the same client. A damage to one of the existing refinery units due to causes originated from the project is dealt with as owner's existing properties. In another case, the project may consist of a new unit inside a petrochemical complex, where different companies own various existing units. In this case, a damage to one of the adjacent properties is treated as third-party property damage. While there is no difference as far as insurance indemnification is concerned, it is important to understand the conceptual difference, because in the negotiation phase, it is necessary to analyze the overall situation in order to come up with the best scheme and make sure that there is no double dipping, or anything left out of the scheme.

SUBCONTRACTOR'S EQUIPMENT

Damages to the construction equipment owned by subcontractors are not covered by the CAR/EAR policy; therefore, it is necessary that subcontractors take out a policy for such damages that may occur during construction.

VEHICLES

These are standard policies taken out by whoever owns and uses a vehicle against damages that may either be caused by the vehicle or happen to damage the vehicle. It works more or less like the domestic insurance policies for private cars.

STATUTORY INSURANCES

These are the ones mostly required by the laws applicable in various countries where the contract execution takes place, mostly relevant to personnel, indemnification, compensation, etc. Policies, like workers' compensation, are sometimes enforced by large international investing owners to be particularly taken out by each party involved, even those parties may already have some coverage in line with the laws of the country where the project is executed. The reason is to ensure that the investing owner is fully protected against any claims and indemnifications in that country.

PRICE AND TERMS OF PAYMENT

The clause should deal with several aspects: Price and its breakdown, currency(-ies), validity, and terms of payment.

LUMP SUM PRICE

The price is expressed as one single amount, usually qualified as fixed and firm for the duration of the contract. A breakdown is normally provided, limited to one or two levels, like (i) price of each single plant if the contract is for more than one and (ii) price for engineering, material supply, and construction. The reason why the price is broken down this way is mainly connected with the terms of payment, which, in a lump sum contract, require the definition of a WBS, of which the two levels mentioned above are just the upper ones. The remaining levels are normally broken down at the level of payment and project control procedures.

Another reason for the breakdown is that in the commercial negotiation phase, the owner believes to have more control over contractor's price if it is broken down into many small components and this, in owner's mind, could help obtaining discounts. When dealing with a complex proposal covering more than one unit, contractors are generally embarrassed by the request of deeper breakdown, because this may later constitute the basis for a request of limiting the supply to only one of the units. In this case, it would be difficult to recover fixed costs that are normally spread among the various packages of work. This is the case, for example, of the cost relevant to the project manager. He/she will work on the project, whether it contains one or two or more units. Therefore, the cost for that position would be originally spread among the units. If the initial request is, say for two units, each of the broken down prices will contain a half of the project manager's cost. Therefore, if the owner requests to remove one of the units from the scope of work, the breakdown would be no longer valid, as half of the project manager's cost would not be recovered. For this reason, contractors' proposals usually qualify the breakdown in various units as "not valid for separate award". If possible, a bottom-up approach is always preferable to a top-down one. In other words, it is safer to offer two units with two separate, individual prices and then a discount for combined award, rather than a single price broken down into two units.

CURRENCY(-IES)

Currency is often a controversial issue if owner and contractor belong to two different economic areas, like Euro and Dollar. The issue is always who takes the risk of the exchange rate. Certainly, contractor is not a gambler and the most international ones prefer not to take such a risk. If forced to accept a price in a currency that is not the currency of the very major portion of the costs, the contractor would then seek protection in the form of financial instruments, which are in any case an additional cost item and therefore, may add to the posted price, tending to push it out of competitiveness. For a contractor, the best solution is probably to make up a basket of currencies, of which composition should mirror the contractor's expenditure plans, in order to limit the risk, for example, asking for payments in Euros for the materials purchased in Europe and local currency for local costs such as construction.

VALIDITY OF PRICE(S)

As said, the lump sum price is generally fixed and firm for the whole duration of the contract. This is possible in places where inflation is under control and more or less predictable. The problem arises when the project is in a country where local prices are not stable. In this case, contingencies or edging contracts are of little or no use and it is preferable to use an escalation formula, i.e., a formula that allows updating the price in accordance with the conditions prevailing at the time of invoicing a given portion of the work.

The simplest method is based on an index published by the local institute of statistics, for example, refer to A4.1:

$$P_i = P_0 * \frac{I_i}{I_0} \qquad\qquad (A4.1)$$

where:

P_i = the portion of price to be invoiced in month "i"
P_0 = the corresponding portion of price, as calculated originally at month "0"
I_i = the value of the price index published for month "i"
I_0 = the same Index, as published for month "0"

In this way, once the amount to be invoiced in month "i" is calculated on the basis of the original price P_0, the use of the formula would return the value P_i to be actually invoiced in that month, recovering the corresponding portion of escalation.

An evolution of this formula is the introduction of a fixed portion, not escalated, as in the following formula:

$$P_i = P_0 * \left(0.20 + 0.80 * \frac{I_i}{I_0} \right) \qquad\qquad (A4.2)$$

The symbols have the same meaning as in the first formula, but in this case only 80% of the price is escalated, while 20% is fixed. This implies higher acceptance of risk by contractor and it is justified by the consideration that the fixed portion represents the contractor's profit, not escalated due to keeping contractor motivated to complete the works as soon as possible.

A further evolution may be a polynomial formula of the type of following:

$$P_i = P_0 * \left(0.20 + 0.15 * \frac{I_i}{I_0} + 0.40 * \frac{ST_i}{ST_0} + 0.25 * \frac{CE_i}{CE_0} \right) \qquad (A4.3)$$

where, in addition to the previous symbols, the others have the following meaning:

ST_i = the price index for steel in month "i"
ST_0 = the price index for steel in month "0"
CE_i = the price index for cement in month "i"
CE_0 = the price index for cement in month "0"

In this case, the factors 0.15, 0.40, and 0.25 reflect a hypothetical cost structure, where services account for 15% (with I_i and I_0 represent consumer price indices), steel for 40% and cement for 25% of the overall Project cost.

The use of a complex polynomial formula brings in more accurate back to back alignment to the actual cost conditions. On the other hand, when using such indices, one should make sure that they are really published and updated within reasonable periods of time. In some countries, indices are not published regularly and the contract should state what to do in this case, to avoid the risk to hold off an invoice for a number of months because of lacking information on the escalation. In these cases, it is preferable to invoice without escalation and cash the principal amount and then invoice the escalation part of the cost once the index value is published. Of course, this mechanism should be clearly spelled out in the contract.

Terms of Payment

If the old saying "a dollar tomorrow is not like a dollar today" is recalled, it is possible to draw a corollary: "There is no price without terms of payment". At every level, be it a proposal or a contract, the posting of a price should always be associated with the terms for its payment. There are many possible ways to structure this clause and its main elements are, as a minimum, the following:

- Down payment, if any; its amount and conditions (e.g., against invoice, bank bond)
- Payment system for services, if based on actual progress measurement, or earned value, or milestones
- Payment system for materials, if based on contractual value for each piece of equipment and bulk materials, obtained from an agreed WBS
- Payment system for construction portion, mostly based on earned value
- Down payment handling, as a portion of it must be deducted from monthly invoices in the same percentage. For instance, if 10% advance payment is received, once the earned value for any given month is determined, only 90% of that value should be invoiced and cashed
- Maximum number of days available to owner for checking and approving earned value statements before contractor submits an invoice
- Maximum number of days for payment of the invoice
- Provision for disputed amounts in an invoice. Generally, in the presence of disputed amounts, it is possible to pay the undisputed portion within the original terms of the invoice and then start a process to clarify the disputed amounts and come to an agreement, so that contractor can submit a revised invoice and cash the balance
- The way that the payment is to be affected; by check or telegraphic transfer, or letter of credit or else

PAYMENT INSTRUMENTS

As said, payments can be affected in different ways, using different payment instruments, like the following:

- By check
- By telegraphic transfer on a bank account indicated by contractor
- By Letter of Credit: These are financial instruments introduced by the Genoese traders around the 14th century, when they traded goods with remote places in the Far East and they were afraid to receive payments in gold or coins, because of the risks of sinking or pirate attack during long return journeys. Therefore, thanks to agreements between banks, they managed to obtain a letter from their Chinese client, requesting a Genoese bank to pay a certain amount on behalf of the Chinese client

Letters od Credits (L/Cs) of modern days have different structures, but the concept is still the same: contractor will receive the payment from his bank in his own country and contractor's bank will in turn receive the money from the owner's bank in the owner's country. The text of the letter of credit must contain detailed provisions as to the payments and the required documentation. Owner and contractor must agree on it at the time of contract negotiation. Therefore, the letter of credit text should be included in an annex to the contract. In addition, the main conditions are often spelled out in the payment clause. A letter of credit may be the following:

- Irrevocable: This means that once issued at day one, it should be accepted by the bank, which should make the payment provided that contractor meets the requirements of the L/C (and the contract), in terms of both substantial and formal requirements.
- Confirmed: This is a condition seldom accepted, because it means that someone has to counter-guarantee the payments, due to untrustworthiness of the issuing bank. Besides being considered as an offense, the counter-guarantee has a cost that should be borne by the owner.
- Revolving, when it is not issued for the full and final amount, but for a lower one and has a mechanism to replenish it while funds are drawn by contractor. This mechanism is devised to limit owner's financial exposure, but it creates some perplexities to the contractor who is always concerned that funds may become unavailable.

FINANCIAL GUARANTEES

When a contract starts, it means that the owner has made a fundamental choice and put its own destiny in the hand of the contractor. For this reason, some cautiousness is required, as the owner needs protection against possible problems along the road. This is obtained through financial guarantees, i.e., instruments issued by a financial institution, stating that under certain conditions owner has the right to automatically cash money to protect its interests. Various types of instruments are analyzed and main financial guarantees that are typically issued on a large LSTK Contract are discussed below.

TYPES OF INSTRUMENTS

Bank Guarantee

The most common instrument used is a bank guarantee, a document issued by contractor's bank to owner, stating that upon owner's request the contractor's bank shall pay certain amounts to the owner. The payment – technically, an "execution" – shall typically be made "at first demand", i.e., contractor has no option to appeal the execution of the guarantee.

A point that is often subject to disagreement in the negotiations is the execution procedure, more specifically, whether owner should give reasons for the execution or not. Contractors accept the concept of first demand and execution without objection, but they request that the owner gives an explanation. It must be noted that many legislations consider the unjustified execution of a guarantee as a criminal offence. Therefore, contractor is entitled to an explanation and may file a recourse in court on the ground of frivolous execution.

In terms of obligation, it is to be reminded that when a company issues a bank guarantee, it assumes a liability. An amount of money must be allocated in the balance sheet to cope with this liability, which will affect the company's credit ranking.

As an indication, the cost of a bank guarantee will depend on many circumstances, but it can be estimated to be between 0.30% and 1.00% of the guaranteed amount per annum.

Insurance Guarantee

Insurance guarantees are similar to bank guarantees as far as their practical use is concerned. In reality, they are more similar to an insurance policy and the risk of losing the insured amount is shifted from the guaranteeing company to the insurer. For this reason, the guarantee does not affect credit rating, but of course, its cost will be considerably higher, close to the range of an insurance premium. Contractors are in favor of this type of guarantee, but project owners seldom accept it, as there is a prejudice about resistance by insurers to the execution of the guarantee.

Standby Letter of Credit

Standby letters of credit are another instrument, more popular in North America than it is in Western Europe or the Middle East. It is essentially a letter of credit issued by contractor in favor of the owner, with the same inter-bank mechanisms discussed for letters of credit. The letter will remain not used for ordinary situations as it is not an instrument for ordinary payments; only in case of necessity, will the owner execute it partially or totally, just like a bank guarantee.

Parent Company Guarantee

A owner may consider contractor's financial situation as too weak to undertake the contract and may therefore request contractor's parent company to back it up by means of a guarantee, called Parent Company Guarantee (PCG). Many contractors belong to larger engineering and construction groups or large industrial conglomerates, or financial institutions and so, have parent companies. Financial capacity of the parent company is certainly higher than its subsidiary, which gives owner confidence against financial risks during execution.

A PCG can be of two types:
- A financial guarantee, which can be executed by the owner in a similar way to a bank guarantee
- An obligation to step in in lieu of the contractor, assuming the obligation to execute the work in case the contractor is no longer able to do it for any reason whatsoever

Performance Bond

A performance bond is a financial guarantee requested by owners from contractors, in order to make sure that contractors will execute the project as set forth in the contract. For the sake of clarity, this is not a financial tool to cover for liabilities in respect of process performances, but it has a different rationale in respect of general project execution.

The bond may be in one of the forms described above (bank guarantee, insurance guarantee, standby letter of credit). Some world-class project owners perform a financial assessment of prospective contractors, to determine whether their financial standings would support a bank guarantee, or otherwise, whether a PCG would be safer. In some extreme cases, both instruments are requested from contractors whose financial standings were deemed by the project owner to be borderline.

A performance bond is valid as long as contractor's liabilities are active under the contract. Usually, its validity period cannot be pre-determined and must be linked with the execution period, in order to align with possible future delays and extensions. At the negotiation table, this is often a matter of discussion. The owner asks for an automatic extension of the validity period to consider possible eventual delays. The contractor may accept, but it would request to limit it to delays for which the responsibility is attributed to the contractor. Moreover, a firm's final expiry date should be provided for all other cases.

Once the contract is closed and contractor's liabilities become void, then the bond amount becomes null and must be returned in original to the contractor. A very important issue here is that in some rare cases, a bond may still constitute a liability even if its value has been reduced to zero and its validity has expired. Therefore, it is absolutely essential that the original of the bond be duly returned to the contractor as soon as it expires.

The amount of the performance bond is established by the owner and is normally between 10% and 20% of the contract price. This amount may be asked to remain valid through the entire validity period, or to get reduced in conjunction with the satisfaction of pre-determined major obligations by the contractor. As a typical example, after mechanical completion, a big portion of contractor's obligations has undeniably been met; therefore, it makes sense to ask for a reduction of the corresponding financial guarantee to the owner. These mechanisms are generally not automatic and must be clearly set forth in the contract.

Advance Payment Bond

In case an advance payment is paid to the contractor, the owner wants to make sure that the money will be used to sustain the contractor's cash flow for the project and is not used to support Contractor's other businesses. Therefore, as a condition for the payment of the advance payment, the owner requests a bank guarantee to cover the advanced amount.

As discussed in the payment clause, owner progressively recovers portions of the advance payment by applying a deduction from contractor's monthly invoices, in the same percentage as the advance payment. In this case, also the advance payment bond may be correspondingly reduced by the amount deducted from invoices. This mechanism can be made automatic, but its use on small and/or short duration contracts is not very common, because that would not be practical.

Retention Bond

Under a contract, the contractor holds certain responsibilities for repairing and making good any defects in the work during all the project phases, from construction to commissioning and testing, up to end of mechanical guarantee period. In addition, the contractor has an obligation to make good the plant in respect of performance guarantees. Many project owners request to apply a retention on the monthly invoices, in order to create a fund that the owner would utilize to carry out the repairs and the make good interventions, in case the contractor refuses to do those. While the concept is widely accepted as being reasonable, its application through a blunt retention of money may not be the right way. This is because it affects the contractor's cash flow and may force the contractor to use bank loans for the project. To avoid this, invoices may be paid in full and the amounts to be deducted would instead be paid to contractor, against the provision of a bank guarantee. In these cases, the guarantee bond may be established at day one with an amount equal to zero and be progressively increased up to an amount equal to the pre-determined percentage of the contract price. This bond can be used by the owner for the purpose mentioned above and the residual value would be zeroed at contract closure

Bid Bond

This bond is used in the proposal phase, not during contract execution. When the owner issues a tender by inviting a certain number of bidders, it starts a process at the end of which one of the bidders is selected for negotiating the contract. The aforementioned process may take a long time and may be quite costly; therefore, if the negotiations fail to generate a contract award, the owner would face serious problems. The selected bidder should normally be happy with this selection, but there may be reasons for pulling back and refusing the award. Whatever the reasons may be, the owner needs protection against this event.

For these reasons, often the owner requests a bank guarantee of a nominal value (the bid bond) to be provided along with the proposal. Such bond can be executed by the owner if the selected bidder refuses to honor its proposal by accepting the contract. The major point of disagreement between an owner and a contractor may be the event that triggers the execution of the bond. The contractor can accept to be penalized for a refusal to execute the contract in accordance with its own proposal, but the request by the owner is often to penalize the bidder if it refuses to execute the contract in accordance with the owner's requests in the RFQ. Such a condition is usually unacceptable to the bidder, who normally puts forward requests for modifications, exceptions and qualifications to the owner's RFQ and for this reason it is definitely not in a position to accept the request, which will be pretty much like a blank check.

EXCLUSIVITY OF REMEDIES

This is a very delicate clause that is often a matter of disagreements in the negotiation phase. Many international contractors would never accept a contract without this clause and this is considered by many as a "walk-away" clause. The point concerns the remedies set forth under the contract in favor of the owner, in case the contractor fails to meet its obligations. As already described, the owner has certain remedies, like applying liquidated damages and demanding make good. Regarding the previous discussion about liquidated damages, it was pointed out that "the damage will be deemed to be finally and totally compensated by the application of that amount and the owner will have no right to ask for any additional compensation". The owner may, however, be tempted to compare the actual damage suffered and the amount recovered in the form of liquidated damages and to conclude that it needs extra compensation that can be obtained by suing the contractor in court. The clause of "exclusivity of remedies" makes it impossible for the owner to do so, as the contractual remedies do suffice to guarantee the owner against all the possible incompliances by the contractor.

There is only a case when the owner can actually seek additional compensation in court and this is if the contractor has committed gross negligence or willful misconduct. The definition of these two conditions must be sought in the legal system of the various countries and it is not possible in this appendix to compare them all. In simple terms,

- Gross negligence means that a blatant mistake is committed, of a nature that a less-than-skilled engineer should never have done.
- Willful misconduct means that something wrong has been deliberately done, knowing that the consequences would be detrimental for the project and with the possible intent to cause any damage to the project.

In either of these conditions, the protection offered to the contractor by the exclusivity of remedies clause fades out and the contractor runs actually into the risk of being taken to court. The damages may be so high in an amount that could even drive it out of business. In fairness, it must be said that these two types of wrong-doing are very difficult to demonstrate in court, in particular, the gross negligence. However, in any case the protection offered by this clause is never given up in order to be fully covered against all kinds of possible adverse situations.

Very importantly, this is one of the few clauses that should always be handled with the support of a lawyer, be it in the negotiation phase or in the contract management phases or – with more reason – in the litigation arena.

INDIRECT AND CONSEQUENTIAL DAMAGES

Here is another clause to be dealt with by seeking legal counsel. It is, for many international contractors another walk-away clause. The remedies contractually available to the owner are meant to compensate for the direct damages caused by the contractor as a result of its incompliance with the contract. There are other damages that the project owner may suffer, like the following:

- Loss of profit: Consider what happens if a large petrochemical complex is built in expectation of a peak of high prices that should make its commercial exploitation viable and then, out of contractor's fault the plant is completed 2 years later, when the peak of products price is over and so the commercial operation would be less viable or even in the loss.

- Loss of contracts: As an example, the project owner could have secured a contract with another entity, for sales of the product produced by the project. The product proves defective, not meeting the specifications required for its commercialization and so the sales contract had to be cancelled.
- Third party's claims: Third parties may suffer damages, whether direct or indirect, because of the contractor's incompliance and they may claim the owner for these damages. Regardless of whether the third party's claim to the owner is grounded, the contractor is alien to the process because this is a consequential damage from the contractor's perspective.

The clause for the above damages should clearly state that neither party is liable to the other for indirect or consequential damages. Such a clause may actually be very long and articulated and the foregoing is a very brief outline of its meaning.

Once again, gross negligence and willful misconduct can play against the faulty party, by making the exclusion inapplicable.

TERMINATION AND SUSPENSION

Suspension and termination are two events that may occur during the life of a contract.

SUSPENSION BY OWNER

Owners have the right to direct contractors for suspending the works for any reason, like lack of permits or authorization, force majeure, etc. In such cases, contractors are obliged to suspend the work, ensuring that the installations are left in safe conditions. Depending on the anticipated duration of the suspension, the contractor may demobilize its project team, or keep it on standby. In either case, the cost will be borne by the owner. Keeping people on standby is expensive; however, trying to re-mobilize personnel after letting them go may even be worse, as the contractor cannot guarantee that the same people can be assigned to the project. If this cannot happen, then the project may face serious disruption. The suspension clause should also state the contractor's right to terminate the contract in case the suspension lasts more than a certain time established in the contract.

TERMINATION FOR OWNER'S CONVENIENCE

At any moment during project execution, the owner may decide to terminate the project. Such a decision may be due to a variety of reasons, some of which are mentioned below:

- The project was being built in view of a certain market condition, but then the market changed radically and the project would become uneconomical.

- A fundamental permit was rejected, without any fault of the contractor.
- Negotiations for a massive financing have failed or.
- A partner in the joint venture withdrew its participation to the investment.

Many other examples can be given, but whatever the case is, the owner does not have to give any justification. The clause gives the owner the right to terminate if it so desires. This is normal because it is implementing the project at its own cost as being the entrepreneur. On the other hand, the owner must protect the interest of other parties and the contractor should not suffer from any such termination. Therefore, in case of termination for convenience the contractor is entitled to receive from the owner:

- Payment for the portions of the work already completed
- Payment for the portions of the work not completed, but in progress
- Coverage for the commitments made by the contractor, for example, purchase orders issued and payments to be affected against future delivery of goods
- Payment for additional costs due to termination
- Payment of a termination fee, to compensate for the loss of profit. The reasoning behind the termination fee is that the contractor had accepted the project, probably by turning others down, with the expectation of a reasonable profit. Termination fee may either be a flat fee, or a progressively decreasing percentage of the contract price, because the more advanced is the project at the time of termination, the more profit the contractor must have already accrued

TERMINATION FOR CONTRACTOR'S FAULT

The reasons that may entitle the owner to terminate the contract for default (or "for cause") are multiple and the contract must spell them out clearly. One case may be a manifest inability to carry on with the project, or the contractor's financial default, or excessive delay accumulated during execution with no signs of a recovery effort. An important point to keep in mind when negotiating the contract is to try avoiding drastic solutions specified in the contract as the default rule, like termination. For example, it would be dangerous to have a statement, like "in case during execution the delay accumulated by contractor reaches six months, the owner shall terminate the contract". Such a clause can be one of those cases when the medicine is more harmful than the disease itself. In fact, there should always be an escape way and more importantly, with a clause like that, the owner would be forced to terminate, as the contractual wording says that it shall, not that it may or it could. Therefore, a healthier clause would say for the above cases of the 6 months delay that "the owner shall request a recovery plan. If the recovery plan is not supplied within a certain number of days and/or the contractor fails to put in place the agreed measures for another pre-established period, then and only then will the owner have the faculty to terminate". This kind of formulation is not softer than the previous one in terms of enforcement of owners' right, but it offers everybody (and therefore the project) a further chance to avoid a traumatic event such as termination.

Regardless of what the reasons are, the termination clause has to deal with consequences and procedures. In this case, the owner should

- Pay contractor for the portions of the work already completed.
- Pay contractor for the portions of the work not completed, but in progress.
- Stand by the commitments made by the contractor, for purchase orders issued requiring payments to be affected against future delivery of goods.
- Not pay any termination fee.
- Back charge the contractor the additional costs resulting from appointment of another contractor to replace the one in default.

TERMINATION BY CONTRACTOR

In most contracts, the contractor is entitled to terminate the contract only in case a suspension lasts for more than a given period of time. Nevertheless, there is often a case which is very contentious during negotiations and it is the contractor's request to terminate the contract for failure by the owner to effect payments. For some international contractors, this is another walk away clause.

DISPUTE RESOLUTION

Elwyn Brooks White, an American writer from last century, said "There is nothing more likely to start disagreement among people or countries than an agreement".

Disputes are inevitable consequences of managing an almost living body, like an agreement or a contract and the contract must contain in itself the rules to resolve the disagreements and the disputes that will possibly arise out of its management.

A dispute arises often when a claim is submitted. The first tentative attempt for resolution is through amicable agreement between the project managers. If they fail, then there may be a second level of amicable resolution by the senior management of both companies. If all these amicable attempts fail, then a real dispute starts and so, the contract has to have the function to set forth its rules. There are fundamentally two avenues opened: Courts of justice or arbitration tribunals, which are mutually exclusive. When negotiating a contract, the parties must choose which one to apply and then the other one will be forever discarded.

COURT OF JUSTICE

If the parties have chosen court of justice, then the procedures will be dictated by the laws applicable in the country where the dispute takes place and there is nothing to add here, except that the process could be very lengthy, also due to most judiciary systems foreseeing one or two degrees of appeal. If instead arbitral procedures are the desired solution, then there are different levels and the contract should state which one to use.

CONCILIATION

The dispute is taken to a conciliator, who is an expert figure, whose purpose is to try to help the parties find a solution, by calling them and inviting them to explain the terms of the problem, relying on the fact that mutual understanding of positions is often the key for a disagreement to dissipate.

MEDIATION

A mediator is a figure similar to a conciliator, but he or she also has the ability and the authority to propose solutions, in order to help creating ground for an agreement. The mediator, just like the conciliator, has no power over the parties; he or she is just a third party trying to help reaching an agreement.

ARBITRATION

The arbitration process is conceptually not dissimilar from a court of justice, in that there is an entity who hears the parties and issues a ruling (an "award"). There are some differences:

- The ruling entity may be a single person (arbitrator) or an arbitration tribunal of three or more members (arbitrators), one of whom acts as the chairman.
- The arbitral award cannot be appealed, the parties accept to submit to the binding award.
- An international agreement (New York Convention) accepted by most states makes the award enforceable without any further ado by the ordinary justice, except for the case when the ordinary judge must issue an executive order to force one of the parties to execute the award (e.g. pay the amount established by the arbitral award).
- Arbitration can take place at any venue, also in a country different from both parties' and different from the one where the project is executed, provided that the parties have so agreed in advance.

In order to make an arbitration procedure effective and enforceable, it is important that the parties agree in the contract the terms of the so-called arbitration agreement:

- Venue of the tribunal: The most common ones in international contracts are London, Paris, Geneva, Stockholm, New York
- Number of arbitrators and procedure for their appointment
- Rules and procedures to be used
- Language of the arbitration, which is independent from the venue
- Agreement to submit to the binding award

Once any of the parties files an arbitration request, the relevant procedure starts. The first step is the appointment of the arbitrators. In case of having only one arbitrator, he or she is jointly selected by the parties. In case of having three or more arbitrators, each party appoints one arbitrator and the appointed arbitrators appoint jointly the chairman. Once the tribunal is in place, it will start analyzing the case, studying the memoires submitted by the parties, chairing expert witnesses' examination and cross-examination and finally issuing an award.

The process is shorter than a court case and international organizations like the International Chamber of Commerce (ICC) have standard timetables that everybody tries to comply with. On the other hand, arbitration is more expensive than court, because the costs of the tribunal are higher (to be borne by the losing party as in court option) and the bill is a quite long one, covering a variety of expenses from arbitrators' fee to fixed tribunal costs, administration fees, living and lodging in a foreign country, etc.

DISPUTE AVOIDANCE BOARD

This is a relatively recent institution, consisting of a permanent board of representatives from both parties plus experts appointed by the parties. Main duty of the board is to monitor the performance of the contract by means of periodical review meetings, with the aim to detect any signs of outbreak of disputes and deal with them as long as they are not critical, before they reach at the stage of a litigation. Although this is not a guarantee of a dispute-free contract performance, it is of a tremendous help because a typical characteristic of all disputes is that they start like a small fire and then, if they are not controlled, they can turn into a major devastating disaster.

Appendix 5
Financing

Scope of this appendix is to provide general knowledge about the mechanisms that govern the funding of a project. This is not intended to be a manual of financial science, as there are many references in libraries or bookstores, which are waiting for interested people to buy and peruse.

This appendix will only provide general information to describe the mechanisms, to the extent that a project manager or an engineer is required to manage relevant project activities. Financial arrangements have an impact on the project risk profile, on the stakeholder matrix as well as on the way the project activities are carried out, influencing procurement of materials and even, engineering.

HOW TO GET MONEY TO "FINANCE A PROJECT"?

When a project owner is planning to build a new project, one of the first question that it should answer is how to get the project funded, how to get the money that is like a lymph to the project. Money should flow to the project in a continuous and reliable way, otherwise it is evident that the project will suffer, slowdown, and eventually starve due to famine.

There are many ways to finance a project, like the following:

- Off balance sheet
- Commercial loans
- Export credit
- Project financing

Each of them has a cost and a different profile of risk for the project owner, because whoever provides money for a project requires certain guarantees and these always mean restrictions on project owner's finances, or assets, or the project itself.

Off Balance Sheet

This is the simplest way to obtain funds, taking them from owner own money. This choice can be implemented with different payment methods, as seen in Appendix 4 – Contractual Clauses, as the money can be directly transferred from owner's account to contractor's account through telegraphic transfer (or Internet banking) or can be made available to contractor through letters of credit. Like in all other methods there are pros and cons.

Certainly, using one's own money limits debt and exposure to financial institutions, improving future cash flow by crossing out passive interests. On the other hand, using one's own funds increases the entrepreneurial risk because the project owner commits its own resources instead of seeking contributions. Probably the worse negative effect of this solution is the financial leverage effect, which in this case plays very negatively against the owner.

Just as a brief reminder, Internal Rate of Return (IRR) measures the financial remuneration to the money spent for the project, regardless of the origin of the funds. The "financial leverage" effect measures the remuneration of the risk capital, i.e., owner equity remuneration. This effect depends on the ratio between equity and debt, in a way that the lower the equity, the higher is its remuneration. In general terms, if there is no debt, then the equity remuneration is equal to the IRR. If debt is started to be introduced into the financial model, it will be seen that equity rate of return will become higher than the internal one and the gap will increase as long as the debt to equity ratio increases.

In conclusion, if a project is financed with owner's own money, an equity remuneration just equal to the IRR will be obtained, giving up the leverage effect, which could have otherwise caused borrowed capitals to be remunerated at a higher rate.

COMMERCIAL LOANS

The scheme of a commercial loan is conceptually very simple: A bank gives the investor the money needed, against payment of an interest. The bank will also require certain guarantees, typically through liens on owner's properties.

The drawback of this scheme is that it is very expensive, as interest rates are at commercial level and the bank has no control over the project risks; in case owner starts repayment at the beginning of commercial operation of the plant, then pre-repayment interests must be applied and this becomes very onerous on the cash flow.

Generally, commercial loans are only used for short periods of time and as a way of transition to cope with the time required by the coming into force of other financial instruments. Let us imagine a project in which everything is ready for the start, a contract has been negotiated and signed and the contractor is ready for mobilization, but nothing can move ahead because there are no funds, as the project financing needs three more months to come into force. In this case, owner may borrow money from a bank to effect a down payment to the contractor and get it to start working. Use of these funds can be continued to affect the first or two of the progress payments, until the funds from project financing arrive and allow the project to start full speed to go ahead. With these funds received at last, the project owner can also repay the bank loan, so limiting the effect of the high interest rate on company's finances.

As far as the contractor is concerned, the use of commercial funds for an interim period is not a major issue per se, but the contractor will always be a bit nervous because the main flow of funds is still on hold. Contractor will only sigh with relief once he knows that the project has no financial encumbrance to go ahead.

EXPORT CREDIT

In the 1970s, the international community devised means to support exporting companies with favorable financial conditions. The idea was to create a set of rules that would allow the administration of exporting countries to provide credit facilities to companies of their country that were in the business of exporting goods and services, including the supply of complete plants on LSTK basis. In 1976, the so-called "Consensus" was created by some OECD Countries, to set up rules for this support; in particular, the aim was avoiding wild competition based on excessively favorable or even dumping conditions. The scheme has been followed since then and today, it is estimated that nearly half a trillion dollars worth of export credit is granted every year, to support the export of goods and services.

The main characteristics of an export credit, from the user's standpoint are: as follows

- Interest rates not less than a set limit, established by tables defined by OECD
- A certain percentage of the value of the credit, as a minimum, should be spent in the exporting country; this value is commonly set at 15%
- Repayment in 6-month installments calculated with a fixed capital portion and a decreasing variable interest portion computed on the residual debt
- Favorable conditions on repayment period, which usually start after the beginning of plant commercial operation phase and can reach a maximum of 10 years for developing countries
- Insurance of the credit

The entities providing the funds are agencies set up by the states that support the consensus and are known as Export Credit Agencies (ECAs). In conjunction with such ECAs, there are insurance companies specialized in the provision of the relevant credit insurance.

It is noteworthy that the insurance coverage provided by the above-mentioned insurance companies will be limited to political risk only, with the exclusion of the industrial risk. Originally, they used to be part of the same ministerial authorities and have little by little gone commercial, becoming fully fledged insurance companies. The insurance premiums can be considerably high, depending on the political stability of the country that imports the goods and services.

A project can be funded by different ECAs, in which case the above conditions will apply to each portion of the financing. Consequently, there may be several packages of purchases in different countries, with different insurance companies and different insurance premiums to be paid.

The most popular ECAs and relevant insurance sections are as follows:

- United States: US Exim Bank
- Japan: Nippon Export and Investment Insurance (NEXI)
- Italy: Sezione Assicurativa del Commercio Estero (SACE)
- U.K.: Export Credits Guarantee Department (ECGD)
- France: Compagnie Française d' Assurance pour le Commerce Extérieur (COFACE)
- Germany: Hermes
- Spain: Compañia Española de Seguro al Crédito a la Exportación (CESCE)

For those who are interested in the technical details of OECD Consensus, the "Arrangement on officially supported Export Credits" is available on the OECD website.

PROJECT FINANCING

This is a scheme that was created to particularly support industries, like power and infrastructures, mostly in regulated sectors. In a nutshell, it consists of making available funds, the repayment of which is guaranteed by the outcome of the project, i.e., sales of electric power, tolls of a highway or a bridge, etc.

The other important characteristic of this financing scheme is that it limits the exposure of the company that receives the funds. For this purpose, generally a dedicated company is created, called a "vehicle company" or a "project company", in the form of a Limited Liability Company (Ltd) with liability limited to its equity; thus, creating a sort of firewall between the project risks and the shareholders. The question that arises now is: If the promoters of the project do not take risks other than their equity, then who takes the risk?

Let us examine the extreme project financing scheme, the so-called No Recourse Project Financing and let us suppose that the project in question is a power plant. In a No Recourse Project Financing scheme, the promoters, who are also called the sponsors, do not assume any risk, beyond the funds they assign to the project company as equity contribution.

The overall scheme is given in Figure A5.1, where the main interfaces are shown; each interface corresponds to one or more contracts to be concluded and managed.

The scheme shown is a very simplified one, as for a complex project it would not be unusual to find as many as 60–70 contracts. The scope of the interfaces is briefly as follows:

* Feedstock supplier, to provide a guaranteed supply of the feedstock required for power production, be it gas, fuel oil, coal, or whatever else
* Product offtaker, to guarantee continuous offtake of the electric power for a minimum number of hours per year. This role clearly must make reference

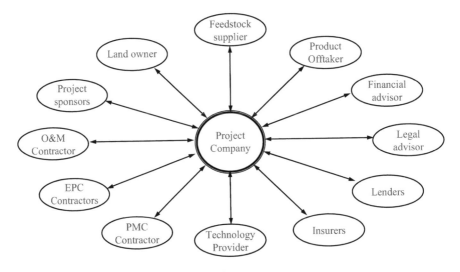

FIGURE A5.1   No recourse project financing scheme.

to the regulatory schemes in place in various countries: The more strictly the sector is regulated, the higher the protection that the offtake agreement can bring to the project

- Financial advisor, required since the very beginning in order to define the scheme, to set up and run the financial model
- Legal advisor, to sort out the complexities of the scheme from the legal point of view, i.e., of the laws in force and the contracting aspects
- Lenders, a bank (the arranger) that undertakes to lend the money to the project company, having the project's output as the sole guarantee. Often the arranger shares the financing with other banks, creating a so called syndicate of banks
- Insurers, to provide insurance protection to the project company in respect of the risks connected with construction and operation of the plant
- Technology provider, in case the plant is based on a licensed technology
- Project Management Contract (PMC), a company selected to manage the EPC construction contract on behalf of the project company
- EPC contractor, the company selected to carry out engineering, procurement, and construction of the plant, under a contract with the project company
- Operation and Maintenance (O&M) contractor, the company selected to operate and maintain the plant, responding to the project company. The O&M contractor's shareholders are often the same as the project company's, but the project financing philosophy requires that the responsibility for O&M be split from the ownership of the plant
- Project sponsors, the shareholders who are linked with the project company by a series of contracts: The investment agreement (among them) and services agreements for the execution of services and secondment of personnel to the project company, which is often created from zero and needs shareholders' resources to operate
- Land owner, the owner of the land (on which the plant is built), who must either sell it or lease it to the project company with a lease contract for the duration required by the lenders

The main characteristic of all the contracts envisaged by this scheme is to relieve the project company and the lenders from as many risks as possible in connection with construction, operation, maintenance, and the commercial operation of the project.

Thus, for example, the offtake agreement shall be construed in a way that the offtaker commits to withdraw and pay for the power produced, even in case that he does not withdraw; as per the scheme known as "take or pay". Likewise, the EPC Contractor will have such liabilities as to tend restoring the result of the financial model even in case of not meeting the guaranteed plant performance.

These two examples show clearly how risky the business of No Recourse Project Financing is for stakeholders, in order to protect the lenders.

The parameter the lenders are looking at when defining viability of the model is the Debt Service Coverage Factor (DSCF), i.e., the ratio of operating margin to the debt service. For example, if the debt service is 100, the lenders want to make sure that the operating margin is at least, say 250. This is to create a buffer margin to "protect" the debt service even in case of major problems to the project, in order to ensure safe return of capitals and interests to the Lenders.

(Debt Service is the amount of money that the borrower shall pay to the lender in order to repay the debt, including principal, interests and costs, or in everyday language the installment).

The scheme had become very popular in the 1990s and beyond, but its popularity started soon to decrease mainly because of the excessive cost structure, due to several factors such as the following:

- Lenders requests for a very safe project scheme, including redundancies of the various systems. To mention an example, lenders can request that, besides a feedstock supply agreement, the project company should also have agreements with third parties in order to secure the supply of an alternative fuel in case the main one becomes unavailable.
- All the stakeholders who have very risky contracts (an example for the offtake and the EPC contracts is given above) are forced to mitigate their own risk by adding contingencies to the prices.
- Legal and financial advisory fees are very high. Moreover, most of them are to be borne before the so-called financial close, i.e. before the lenders start to deliver funds to the project. This means that the sponsors have to bear the cost of several tens of millions of dollars without having the certainty that the project will see the light.

It was estimated that in many No Recourse Project Financing schemes, the final project cost was easily 30% higher than it would have been without that scheme.

Appendix 6
Risk Management

When discussing risk management, it is important to ask oneself: What is a risk? The answer is not obvious, as the word "risk" has a meaning in everyday life and a slightly different one in project management.

Let us start with a question: Is it riskier for a blindfolded man to cross a highway where very few cars move at 140 km/h, or to cross a city street where a bumper-to-bumper traffic tailback allows a speed of only 20 km/h? The common sense would tell that the highway is something to stay away from as a pedestrian, because the consequences of being run over by a car at that speed are certainly fatal. Let us consider the situation in a rigorous way. If the traffic on the highway is not intense and fast and that allows enough time between one car and the next, then the probabilities to be run over are very low. Yet the consequences are catastrophic. On the other hand, trying to cross a city street blindfolded while cars are hardly moving at 20 km/h, there is more probability to be run over, but the consequences will be much less severe.

From this example, it is seen that the parameters for a correct risk analysis are two: (1) "The probability that a certain negative event happens" and (2) "The consequences it may bring about". Well, in project management language, a risk is the combination of two factors: Probabilities and consequences; therefore, the risk that the two blindfolded eccentric persons are running may be comparable.

Risk management is an art, because the matter of risk is not mathematically straightforward. For example, by multiplying probability and consequence, a definite measurement of the risk cannot be obtained, for the simple reason that a risk is just a statistical concept. If a coin is thrown up in the air 100 times, it would almost certainly end up with the final result of getting to very close to 50%/50%. The foregoing does not obviously mean that betting, for instance, on "tails" simply because last time it was "head", is the right decision. Hence, risk management is the art of supporting management decisions with probabilistic data, but in the end the person who decides will always do it with a probability to repent later.

TYPES OF RISK

Regarding projects, there are basically two types of risk:

ANTICIPATED RISKS

These are negative events that are very likely – if not certain – to happen, but their extent is impossible to predict, therefore, the consequence on the project are difficult to estimate. Below are some examples:

- Fluctuation of currency exchange rates
- Incomplete or poor technical definition
- Third-Party approvals, sometimes a lengthy process, sometimes unpredictable with regard to issues that may emerge
- Job site conditions different from anticipated, soil sampling in the proposal phase may lack accuracy
- Escalation and market risk (though everybody knows that prices will go up, but not by how much)

Non-anticipated Risks

These are events that may or may not take place, but it would be safe to analyze them in the context of project. Some examples are as follows:

- Changes in law
- Changes in fiscal regime
- Restrictions to import or work visas
- Embargo
- War, riot, terrorism
- Anomalous process for local permits/authorizations
- Restrictions to exporting currencies or profits
- Financial default of local partner
- Financial default of the foreign country where the plant is being built

RISK STRATEGIES

Various approaches about risks are possible once a risk is identified.

Mitigation

Some measures, which can reduce the probabilities and/or the impact, must be taken and as a consequence, the risk may become acceptable. The mitigations may be of different types, but they have almost always something in common: They have a cost, which is accepted in order to reduce the probability of suffering bigger consequences.

For the sake of analysis, let us classify the mitigation measures as follows:

- Physical, for which probable measures are as follows:
  - Use nonflammable materials to reduce the probabilities of a fire.
  - Install sprinklers to reduce the impact of a fire.
  - Install an anti-intrusion system to prevent burglars from breaking into.
  - Lay a napkin on your knees when eating at a restaurant, in order to protect your trousers.
- Procedural, that can be mitigated by the following:
  - Smoking prohibition in hazardous areas

- – Quality assurance and quality control, to improve chances of control over the products and the processes
- Psychological, to be coped with the following:
  - – Operators training on safety matters
  - – Sanctioning wrong behaviors, like fining operators found without Personal Protective Equipment (Helmet, safety shoes, gloves, goggles, etc.)
- Execution related, which may be overcome by the following measures:
  - – Split a piece of work into two, e.g., dividing the scope of work between two subcontractors with a commitment from both to take over the other part, if necessary. In such a way, if one of them fails to perform or goes bankruptcy, the replacement is readily available without major disruption to the Project.
- Economic, that can be handled by the following:
  - – Adding contingencies so that if the event takes place, its impact is at least partially covered by an amount allocated in the budget

Once again, almost all the above measures have a cost, which ranges from the physical cost of sprinklers, to giving up economy of scale when appointing two contractors in lieu of one; from training cost to the risk to lose a job for adding too much contingency in the proposal.

AVOIDANCE

It is crucial to avoid getting into a situation in which the risk could become real and not controllable.

As highlighted in the Appendix 4 – Contractual Clauses, many international companies would rather lose a contract than taking it with unacceptable conditions such as indirect and consequential damages, or no limitation of liabilities.

TRANSFER

A risk can be transferred onto another party, which is in a better position for taking it. For example,

- Transfer the risk to an insurance company, through payment of a premium.
- Transfer the risk to a contractual counterpart (e.g., from contractor to owner) through negotiation of contractual clauses. For instance, transfer to the owner the risks connected with late release of local permits.

ACCEPTANCE

This is the case with risks that have very low probabilities, or very low impact, or both. In these events, the risk can be simply accepted as it is, proceeding without further action. It must be noted that acceptability can be intrinsic with the risk, or may be the result of mitigations or transfer; in these cases, the project manager would accept the risk as modified by the mitigation measures.

RISK ANALYSIS

Qualitative Analysis

This is a very basic, but most commonly followed practice for risk analysis. It is conceptually an easy exercise; however, the same cannot be said of its assumptions and conclusions, which are often subjective and questionable. This appears normal in a discipline, like risk management, which is based on statistical data, with no absolute certainty.

On the other hand, the qualitative risk analysis can as well be a complex exercise, but a simplified version in a proposal phase will be described here for the sake of explanation.

Identifying the Risks

The first step consists of identifying the risks and the better way to do so is in a brainstorming session, where a facilitator would encourage attendants to come up with all the risks they can think of. Many companies have its own format, with which all the possible risks are identified and listed as grouped by areas (e.g., technology risks, engineering risks, market risks, construction risks, contractual risks). Such a format is usually the result of many analysis sessions done in the past, with new risks added to the standards, whenever identified. The use of fully populated standard lists is very important, because it is much safer specifying a certain risk item as "not applicable" rather than omitting or overlooking it.

Classifying the Risks

The second step is the classification of risks based on their probability of occurrence, for example,

- Low, Medium, and High, i.e., L, M, H or
- Very Low, Low, Medium, High and Very High, i.e., LL (Low Low), L, M, H, HH (High High)

For a general use, L-M-H classification is usually enough.

Then, a similar classification is carried out on the consequence side, leading to the same categories, say L-M-H (Low impact, Medium impact, High impact). The categorization exercise is also better done in a team, with the help of a facilitator. During the exercise, reasons for the classification must be given, so that the discussion about risks is intimately linked with the analysis of project dynamics. At this point, a criterion is needed and a first screening takes place between the acceptable risks and the non-acceptable ones. The criterion must be based on an analysis of the combination of probabilities and consequences. For example, a risk with high probabilities and low impact would be an "H-L". Acceptability criteria depend on each company, for instance, it can be stated that L-L, L-M, and M-L are acceptable; the others need to be mitigated.

Identifying Mitigation Measures

Next step is the identification of mitigation measures; this step is critical and it requires the combined effort of many groups, technical, project management and legal. Each mitigation action should be associated with a rough estimate of its costs. There is no point in brilliantly mitigating all project risks, only to realize that our price has skyrocketed.

Iteration Process

At this instant, an iteration process should start and a new analysis of probabilities and impacts should be carried out on the new basis. Normally, the ranking of most risks should be downgraded and from this effort, a new shorter list of non-acceptable risks can be drawn. By repeating the entire process, the final result often shapes up in a way that most risks turn out to be acceptable – at the expense of the project budget – while a few may have remained as not acceptable. Arriving to this step may be seen as a waste of time, since at the end of the day, there are still risks that cannot be accepted. In reality, a great result is achieved, as reaching a position to provide the senior management a statement of risks, indicating which ones are still unacceptable, despite the mitigation measures considered. The senior management has now the elements it needs to make an informed decision based on project budget (including the cost of mitigations) and risk profile. Without this complete process, it would be unfair and not effective to bring the problem to the highest levels in the company with no elements to decide. Needless to say, senior people will have to make a management decision, also based on other factors, like the interest of the company in the project and the chances of securing the award.

A typical form of qualitative risk analysis is provided in Figure A6.1.

QUANTITATIVE ANALYSIS

There are more sophisticated techniques than the qualitative analysis described above. The main purpose is a more accurate evaluation of the contingencies needed to protect the company against given risks. These techniques are time and resource consuming and so, they are only used in special cases, when the level of contingencies can be critical for the budget, for instance, in a proposal for which a tight competition is anticipated.

Let us refer to a specific risk, which may lead to overrunning the budget, notwithstanding the contingencies provided in it. Clearly, the higher the contingencies, the lower the probabilities to exceed the budget. It is possible to sketch a curve representing the probabilities for overrunning the budget versus the amount of contingencies added to the base cost. The way the curve is constructed is outside of the scope of this appendix, just noting only that the method followed is linked to the Monte Carlo algorithm, a way to generate numbers in a random way. The rationale is to perform an extremely high number of simulations of a real situation, in which occurrence of negative events is often random or unpredictable.

The result is the typical curve represented in Figure A6.2. If no contingency is added at all, then the probability to exceed the budget will be maximum.

Proposal for refinery in Magnolia

Fester Consulting co.

Qualitative risk analysis

Risk no.	Description	Consequence	Before mitigation		Remarks	Mitigation	After mitigation		Conclusion	Extra cost
			Probability	Impact			Probability	Impact		
1	The tender documents are not very accurate	Risk of under estimating, budget overrun	H	H	Price must be lump sum	Add contingencies to the price	H	L	Acceptable	Yes
2	The client has bad reputation for payments	Financial exposure	H	H	Per ITB, payments are by telegraphic transfer	Request letter of credit	L	H	Acceptable	No
3	The project site is dangerous for civil war	Threat to our people's life	H	H		Build our own camp Hire security contractor	M	H	Decision by senior management	Yes
4	Smaller bidders than us may be invited	They will be more aggressive. Unfair competition	M	H		Suggest to client that we would not bid	L	H	Acceptable	No
5	We are not very familiar with the technology	Design inaccuracies resulting in redoing, and failing performance test	H	H	In the medium term, we are anyway interested to develop this field	Hire expert engineers	L	M	Acceptable	Yes
6	Lack of good local construction contractors	Inability to build the plant, quality problems	H	H	High pressure, alloy materials	Bring international contractors	L	L	Acceptable	Yes
7	Restrictions on import of materials - local market protected	Excessive pricing by locals	M	L	Same problem applies to all bidders	Request prices in the bid phase. If possible, make pre bid agreements	L	L	Acceptable	No
8	Lack of good local engineering support	Difficulties for local surveys, as built drawings	M	M	Same problem applies to all bidders	Open our own local engineering company	L	L	Acceptable	Yes
9	Difficult to obtain work visas	Delay in assignment of personnel to site	H	M	Some countries are black-listed by Magnolia	Plan ahead Appoint visa agent	M	L	Acceptable	Yes

FIGURE A6.1   Typical qualitative risk analysis form.

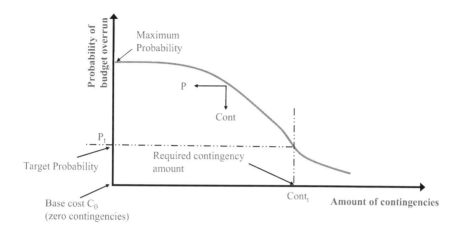

FIGURE A6.2 Overrun probabilities versus contingencies.

Every generic point on the curve tells that, by adding an amount equal to "Cont" to the base cost, the probability will gradually decrease and can be read on the vertical axis as "P". The curve is asymptotically tending to zero, reflecting the fact that, mathematically the chances of an overrun are never an absolute null.

In order to use the above curve, the start should be by stating what probability of overrun is to be accepted (Target Probability, Pt) and consequently, the amount of contingency required can be read from the figure. Often, a balance is necessary between budget security and chances to get the job and some compromise on the chance of overrunning may need to be accepted.

This result can only be used by senior management to make informed decisions and is just a bit more accurate than the qualitative risk analysis described above. Moreover, an important limitation of this method is the accuracy of the curve and the way it is constructed. For all these methods based on mathematical models, it should always be kept in mind the old expression "garbage in, garbage out".

RISK ALLOCATION

One of the ways to define a contract is "a means of managing project risks". The contract should be conceived in a way to provide solid ground for a sound risk management and in doing so, some principles as outlined below must be respected:

- A valid risk management strategy does NOT consist of allocating all the risk on one side, pretending it will be solely a problem of the other side.
- Every risk should be taken by the contract party who has better chances to control it. If not, the risk is not optimized.
- A valid risk sharing leads to global overall risk minimization for the entire project.
- Every risk should have a mitigation measure associated with it.

The importance of these concepts can be conceived better with an example. Let us consider a lump sum contract negotiation phase and the discussions taking places around a certain risk in connection with local permitting. In these cases, the owner often tries to push this risk to contractor's side, while the contractor objects that it is hardly in a position to control the local permitting process, apart of course from ensuring that the documentation provided is correct, which is its undisputed obligation. If as a result of the negotiation, the contractor is forced to accept the risk, then it will also be forced to add some contingencies to the price, to be covered in case something goes wrong with permits. Having said this, let us suppose that during the performance of the project, the event under discussion does really materialize, causing an economic damage to contractor. In this case, contractor will be (at least partially) covered by the contingency.

If instead, the risk does not materialize and everything goes smoothly, the contingency is not used and it will become an extra revenue for the contractor. In both cases, the owner gets into the situation that it has virtually spent that amount of money when the corresponding contingency is approved. Should the owner have accepted to keep that risk for itself, it might have had the chance to save that amount. Moreover, probably the owner would have better control on local permitting than a foreign contractor and this would have led to a lower quantity of contingencies in the price; resulting in both cases to have the owner spending less money.

Appendix 7
Joint Ventures

Associating two or more companies' interests to achieve a goal is a market trend which is very common in today's world. On the global market, it is not unusual that breaking news often emerge on long-term associations, merges and acquisitions, companies that change ownership or the controlling group.

The scope of this appendix is not to discuss such form of association, but rather the short or medium-term formation practices deployed in the engineering and construction industry to execute projects.

REASONS FOR FORMING AN ASSOCIATION

There may be various reasons leading to the decision of associating with another company to jointly bid and, if awarded, jointly execute a project. The most common reasons are as follows:

PROJECT SIZE

In the present market, the economy of scale often suggests project owners to launch large and very large projects, often called mega-projects. These are the projects having investment costs in the range of the tens of billions of Dollars. It should not be a big surprise if a project worth 100 billion US dollars materialized in some place around the world. Under these circumstances, there is probably no company in the world that can undertake such a monster project, mainly due to the following:

Resources

By using rules of the thumb, a 20-billion-dollar project (this may occur mostly in oil and gas) would require something in the range of 30–40 million home office man-hours. If a span of 2 years is assumed for the home office activities, this would lead to a requirement of an average of 8,000–10,000 employees, with a peak of maybe 20,000. It is very unlikely that any company in the world can have this amount of resources and moreover, commit such level of resources to a single project.

Financial Risk

Contractual liabilities may easily be in the range of 2–4 billion and a risk of this magnitude would be seen as a killer for any company in the world. Furthermore, a 1% error in the cost estimate – and 1% is absolutely a normal possibility to occur – would cause a damage of 200 million.

Bonds

If a performance bond and an advance payment bond are required, this may easily mean an exposure of 3–4 billion, which is likely to saturate the bond giving capacity of most companies, including the very large ones.

Even if a company possessed the resources and the financial/bond providing capacity, the question remains whether that company would be ready to put all its eggs in the same basket. Project owners are well aware of these limitations, so that when they are planning a mega-project, they often conceive it since the beginning in a way that it can be split in several EPC packages. Each one would then have a size that would be affordable to many contractors. In this way, there would be more companies interested in the tender and competition would be ensured.

Catch Opportunities

Commercial strategies may push a contractor to try to penetrate a new market by bidding and then executing projects. There are many obstacles to this strategy, including lack of knowledge of local market and local commercial practices, but also there may be reluctance by large project owners for political reasons or current trends in policies to "buy local".

One of the ways to overcome these obstacles would be to associate with a local contractor to jointly bid for such a job and execute it if the bid is successful. Of course, the local entity is aware of the constraints discussed above and inevitably, there will be a price to be paid by the foreign company, like a sort of entry ticket. This may include unfavorable share of profits and/or risks, particular extra efforts demanded by the local partner to facilitate its entrance into other markets and so on.

TYPES OF ASSOCIATION

There are many types of association. Fundamentally, they can be divided into two groups:

1. Individual responsibility or
2. Shared responsibility

Individual Responsibility Associations

Individual responsibility joint ventures are those in which the partners hold sole responsibility over their part of the job, in terms of price and project execution performance.

Consortium

A consortium is the most typical individual responsibility joint venture. In many countries, it is called a temporary association or a temporary grouping; this is the case, for example, of Italy, France, and Spain. The English word, consortium originates from Latin "consors", meaning "joint destiny" and in reality, it seems inadequate to describe this type of association in which the destinies of the partners are not really integrated. The following is an outline for the typical features of a consortium:

- One of the parties is the leader, who receives a mandate of representation by the other parties. This is why in many legislations the leader is called the mandatory, while the parties who confer a mandate upon the leader are the mandators. The mandate has generally very low compass and it is normally limited to a nominal leadership with administrative functions only, but almost no decision power at all.
- The scope of work is split in such a way that physical portions of the project can be attributed to each member. For example,
  - different units in a multi-unit project (vertical split)
  - process units and utilities (vertical split)
  - engineering and supply of materials versus construction (horizontal split)
- Each member is solely responsible of its portion of the price, having carried out its independent cost estimate and pricing process. The price formation is therefore a bottom-up process, as the price to the client is the sum of the individual prices.
- Each member is solely responsible for the execution of its portion of the scope of work, using its own resources.
- Each of the members issues its own invoices to the owner and sends them to the leader. The leader collects the invoices and forwards them to the owner for processing and payment. The owner effects the payments to each of the members separately.

There is one tricky point about consortia, as most project owners require that the consortium members assume joint and several responsibilities toward the owner for the project execution. This means that inability by one member to execute its portion properly would trigger intervention by other members to step in and restore correct execution. Another area where this could be a problem – considering the nature of the consortium – is the joint and several responsibilities for liquidated damages, as the same are inherent to a given portion of the problem, where in reality the responsibility is with one of the members only.

The answer to owner's such request is often a provision in the consortium's by-laws to the effect that, notwithstanding the main contract stating differently, each of the members remains exclusively responsible for its own portion only. This provision works pretty well for "ordinary" problems, e.g., if there are liquidated damages, or if there is a budget overrun; in these cases, the faulty member will honor its individual responsibility on behalf of the consortium and the owner will be satisfied. The real problem arises in very serious cases, like inability of a member to continue to execute the job, bankruptcy, etc. In this case, the owner should require that the other member(s) take over the portion of work of the defaulting one and this could be a problem if the two members (the defaulting and the surviving one) are of different type, e.g., one is an engineering contractor, and the other one a construction company.

SHARED RESPONSIBILITY ASSOCIATIONS

In this type of association, there is a real share of responsibilities, in terms of price and contractual obligations.

True Joint Venture

The typical form of shared responsibility association is the Joint Venture. The last two words are intentionally capitalized, to stress the difference between this specific form of association and the generic term used for just any association. In common practice, it is customary to refer to the specific form as a True Joint Venture. The confusion in the terms is still there, but it can be overcome. Another issue that could generate confusion, is that in other systems, in other languages the word used for a True Joint Venture is consortium. If we remember the Latin origin, the word consortium seems more adequate to describe this form of cooperation, but in English it designates a different form of association. We should pay attention to the terms used when dealing with this issue.

A True Joint Venture operates in a similar way with a joint stock company, as there is a single scope of work and the responsibility is shared between the members in proportion to their financial participation. Below is the main characteristics of a True Joint Venture:

- It is a legal entity with its administrators, its by-laws and management bodies.
- Each member has quotes, which represent the percentages based on which profits or losses will be shared, as well as the contractual liabilities in the joint venture.
- One of the members is the leader of the joint venture. Leadership does not confer any real power of representation beyond what is stated in the by-laws.
- It has fiscal obligations of its own and has to submit a financial statement for each year of business.
- Vis-à-vis the owner there is a single price and a single point responsibility.
- Portions of the project scope can be sublet to third parties, or even to the members. In this way, part of the common scope becomes individual scope, but vis-à-vis the owner the responsibility remains with the joint venture.
- Personnel who manages the common scope can be either seconded by each member, or hired externally. In case of secondment, there will be a service agreement between the joint venture and each of its members, to rule the secondment process.
- There can be mixed management bodies, typically a steering committee and a project directorate.
- The steering committee is formed by representatives of the members and has the power to make decision at the highest level. For example,
    - Regarding decisions on award of large purchase orders or subcontracts
    - Giving guidelines to the project director as to the execution of the project
    - Intervening to rule out on internal disputes that could not be solved at the project director level

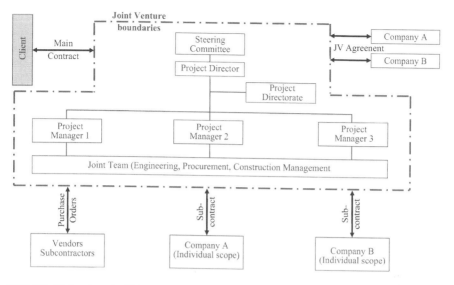

FIGURE A7.1   A typical True Joint Venture scheme.

- The project directorate is a conventional project management team, in which representatives of each member cover various functions. In this way, each of the members is satisfied to have ears and eyes in the most vital positions of the team and keep some control on the joint operations.
- The project director is generally appointed by the leader and has power to direct the Joint Venture personnel (including the seconded personnel), irrespective of which member they come from. Also, the project director has some powers of dispute resolution and can rule on internal contentious matters of small or medium impact on the members.
- Internal dispute resolution process is one of the most delicate aspects. Normally, none of the members has the power to make ultimate decision on any issue. On the other hand, it is not imaginable that each and every decision will need to be taken by the steering committee, or perhaps be taken to an arbitration tribunal. The by-laws, in this respect, must provide a reasonable equilibrium between power of control by each member and the need to operate, which requires decisions to be made by a manager in an effective and expeditious manner.

Figure A7.1 shows a typical True Joint Venture scheme. Part of the scope is sublet to the members A and B, while the rest is procured on the market. As it can be seen, the way a True Joint Venture works is very similar to a joint stock company, with the difference being that it is formed in relation to a specific project. Once the project is over, the joint venture is dissolved.

PROFESSIONAL ASPECTS

Working in any joint venture environment is one of the most challenging experiences for a manager. When working for a company, there is a set of procedures and internal rules to observe. Above all, there is a set of core values which are the basis of the mission of the company. These values are duly absorbed by managers as a part of their induction process in the company and remain in their personal profile to guide their everyday professional life.

When the manager decides to leave the company and work for another employer, the process will start again and he or she will need to learn and assimilate the values of the new company. After a necessary period of acclimatization, the new values will be absorbed and he/she will be ready to start operating under new internal standards and guidelines.

The situation is a bit different when the manager is pushed into a joint venture environment, as there are two or more members, each with its own set of rules and procedures. In this case, the first question is: Whose rules apply? The answer must come from the joint venture management, for example, through a working group to come up with the most appropriate set of rules. So, here is a first implication for the manager: Working to "joint" rules is a professional challenge that requires flexibility and intellectual honesty. Rigidity is not the solution, creativity is most welcome. But the challenge is not limited to the issue of rules, since objectives and strategies are also to be cared for. When a joint venture is created, inevitably, the interest of the members will coincide to a good extent, but they will never be exactly the same. There will always be some room left for disagreement, or for divergent interests. So, the managers working for the joint venture may be exposed to some degree of professional conflict.

Experience shows that the best medicine for these situations is to behave in a professional way. It should always be kept in mind why the manager is in that position and what he/she is called to do. Loyalty goes to the joint venture, as its interests are superior to those of the partners, so one is supposed to work to the benefit of the joint venture, in accordance with its objectives and strategies. This is sometimes easier to say than to do, but it works at least for everyday issues. If a bigger issue comes up, involving a conflict between the joint venture members, then the problem is no longer limited to the manager's professional behavior, but it will have to be solved at a higher level. What matters really is to keep cold blood and clear vision, in a transparent way.

Working in a joint environment forces people to look critically into their own job, like an actor who studies the character he has to interpret on stage and this effort leads to professional growth.

One last point is concerned with the criteria of choice whenever it is needed to look for the best partner to jointly tackle and execute a project. It is important that the structure of the joint venture reflect the nature of the members. An engineering company and a construction company should not become equal partners in a joint venture, in particular, in a True Joint Venture. The mindset and the cultural approach to business is different by each and the conflict will arise sooner or later. In a True Joint Venture, an engineering contractor should partner with another engineering contractor. If an association between an engineering – or an EPC – contractor and a construction company is necessary, then it should better be on a consortium basis, so that one will substantially be held responsible for engineering and the other for construction.

Appendix 8
Transportation, Custom Clearance, INCOTERMS

Transportation is an often-underestimated source of trouble in a project, besides it takes a sizeable portion of the overall delivery time, as well as a non-negligible portion of the budget. Concerning today's Engineering, Procurement, Construction (EPC) projects, for which time of completion is more and more tight, spending 1 month to transport goods from a manufacturer's premises to a remote jobsite is really long and the whole operation for transportation and custom clearance must be dealt with care and in a time conscious way. This is not an easy task, as it is an area where many players have their own distinct roles.

ENTITIES INVOLVED IN PROJECT TRANSPORTATION

First of all, let us analyze the general steps that it takes to transport a cargo from a manufacturer's facility to a jobsite.

FROM WORKSHOP TO PORT OF ORIGIN

This includes packaging, which is generally taken care by the manufacturer and actual transportation by road or train or inland waterways, or often through a combination of all those. Depending on the purchasing conditions, this can be done by the manufacturer, who delivers its goods to the port. In other cases, the delivery point in the purchase order is manufacturer's facility, and the contractor takes care of transportation from there on. This scheme is called "ex-works" as described in later sections of this appendix.

LOADING ON A SHIP

This is an operation that can be done by the port authority through its own craning facilities – often managed by a dedicated contractor – or by the ship itself, if the vessel is equipped with cranes.

OCEAN FREIGHT

In some countries, ocean freight must be done by a national flag sea carrier while in most countries there is freedom to select a carrier among those running that line. Restrictions as above are more likely to apply if the project is for a state-owned project owner and/or the project enjoys custom or taxation exemptions.

When the goods are accepted by the vessel's commander for loading aboard, there are one or more sets of original documents that will accompany the cargo up to final destination and are known as the "shipping documents":

- The packing list, a detailed description of the goods transported, including the content of casings
- The bill of lading, a certificate of takeover of the goods by the vessel, with a summary description of the same
- The certificates of origin of all the equipment and materials shipped. This is required by the owner for control purpose, as well as for custom clearance and in some cases, to cope with constraints imposed by project financing
- The invoice prepared by the manufacturer or supplier for the equipment and materials. Such invoice, along with the other shipping documents, will have to be certified by the owner upon contractual delivery and used by manufacturer/supplier to cash the payment. The invoice is the primary document within the set of the shipping documents

TRANSSHIPMENT IF REQUIRED

For heavy lift or extra size cargoes, it is often more convenient to call local private berthing facilities near the jobsite as reaching the jobsite by road from commercial ports may be risky or not feasible. Transportation in such a case is often done on Roll-On-Roll-Off (Ro-Ro) basis, whereby the equipment is loaded to the ship without removing it from the special truck or lowboy on which it lays. The kind of ships that can allow this delivery mode are like the ferries that transport holiday cars and commercial trucks to vacation sites. A big gate opens on either side of the ship and the vehicles can leave by their own means. The truck carrying the equipment then pulls out in the same way at the arrival berth and the same truck reaches the jobsite for final delivery. This is a much more convenient way to deliver very heavy and/or very large pieces, like packaged units, or process modules. The only problem is that Ro-Ro ships are not designed for long sea fares and are normally operating on short range routes. Moreover, the few large ocean fare Ro-Ro vessels are very expensive to be used. In these cases, the equipment is transported from the port of origin by conventional sea freight, usually laying on deck, as opposed to staying in the hold. When the vessel reaches at a commercial port located in a short range of the final destination, the large equipment is downloaded onto a lowboy and then can be loaded into a Ro-Ro vessel for final delivery to the local berth.

Another reason why transshipment is sometimes required, is that small local ports are sometimes called by few sea lines, therefore the delivery time becomes much longer due to waiting for a vessel that goes from port of origin to port of final destination. In these cases, it may be much faster to transport the materials to a large harbor in the region and from there, to transship onto a local vessel for the final destination. One important point to keep in mind is that, if payment is to be made by letter of credit, then the letter must specify "transshipment allowed"; otherwise, there may be problem in case the bill of lading is issued for a vessel and delivery is made by another one.

Unloading at the Port of Destination

Apart from the case of Ro-Ro vessels, unloading of a cargo from a vessel can be done by the same vessel if it has the proper cranes, if not, by the port facilities. Usually, unloading facilities place the goods on trucks for transportation to the final delivery point.

Custom Clearance, or Transportation to a Custom-Secure Yard

If the goods pass the borders of a country and enter into another one, custom clearance and payment of custom duties are required. In this regard, European Union (EU) is regarded to be a single state and the above does not apply among the countries in the EU.

In many cases, custom clearance operations are carried out upon unloading. Then, the goods are free to proceed straight to the jobsite. In some cases, the shipping documents may lack some information and so the custom officers may not allow the goods to leave the port area until the paperwork is made good. The materials are then stored in a special secluded yard under the custody of custom organization. This yard may be at the port, but it may also be at jobsite, at an area duly fenced and equipped with housing facilities for exclusive use of the custom officers. Once the paperwork is corrected, the contractor is authorized to take the material out of the custom yard and use it in the works.

Custom clearance is an extremely delicate operation, because incorrect shipping documents can trigger loss of days, if not weeks and this may be very detrimental for the project schedule.

Transportation from the Port/Custom-Secure Yard to the Jobsite

Once the goods are custom-free, they can be taken to the jobsite. This requires another transportation contractor that will be responsible for the delivery to the jobsite unless the scope of either the supplier or the previous transportation contractor includes such in-land transportation.

Unloading and Final Delivery

Once the materials are ready to be delivered to the contractor at the jobsite, the handing over occurs at one or more of the following locations.

Storage Yard at Jobsite

This is the case for pieces of equipment, large-size piping materials, and any material that can be stored in the open air without risking deterioration. The equipment and materials will be taken from the yard for incorporation into the works, when required, according to the construction schedule. For storing at such lay-down areas, it is customary to fence the area just for marking the borders and discouraging unauthorized access, though the height of such wire fencing is often not higher than one meter. The materials are located on timber planks in order not to get soiled.

Warehouse

Here the material it is stored on shelves or similar structures. This is the case for materials that would either be damaged or lost if stored outdoors, like instruments, special valves, computers, bolts and nuts, gaskets, small-size piping materials, etc. For some special items, it may even be necessary to store in an air-conditioned area.

Directly on its Foundations at Plant Site

This is usually the case for very large equipment, for which it would be too expensive to unload at a lay-down area and then load again for moving to the location of installation. Large modules or, either heavy or large equipment delivered by Ro-Ro vessels are dealt with in this way. A direct requirement for this approach is that site works must be so carefully planned and executed that the foundations should have been made ready for installing the equipment as soon as it arrives at the jobsite.

Unloading and delivery to the jobsite as above is generally done by a handling contractor, specifically appointed by the contractor. Mechanical erection subcontractor can often do the same.

DOOR-TO-DOOR TRANSPORTATION

From the description above, it is clear that the overall transportation sector has a large number of players involved and therefore, there are many possible organizational schemes to manage the whole bundle of related activities.

Certainly, one possibility is for the contractor to take care of each and every segments involved, coordinating the various players and contractors. It should be noted here that it is not always the contractor who is responsible for handling transportation and custom clearance. In projects other than EPC Lump-Sum Turn-Key (LSTK), such duties and responsibilities may as well be borne by the owners. As usual, there is a balance between "do it yourself" and "sublet it". One important consideration to be kept in mind is the seamless continuity of responsibility. The do-it-yourself formula allows probably to save some money because there are no intermediaries between the contractor and the various players involved. The risk is the responsibility gap and this can be critical, if it is imagined, for example, that late delivery to the port of origin may imply missing the sea vessel and having to wait for the next one with wishes that there is space available on that next vessel.

Moreover, doing it in-house means that certain resources are needed and this means additional staff and so fixed costs that must be factored in when comparing alternatives.

Probably the best – and the most common – solution is to employ a forwarding agent to take care of the whole transportation on door-to-door basis. He/she will coordinate among the various players, such as:

- Manufacturers (to pick up the goods at the workshop)
- Various local transportation companies (from warehouse to the port of origin and from port of arrival to jobsite)

- The sea freight carrier
- The Ro-Ro ship-owner if necessary and
- The custom operations

Of course, the forwarding agent will charge a fee and make a profit, but in most cases this additional expense is well justified by the effectiveness of the solution. A secondary choice is whether the contracts with actual transportation companies are issued by the forwarding agent or by the contractor. The invoicing procedure will be defined accordingly, i.e., whether contractor will pay for the carrier and other transportation contractors directly, or the payment will be made to the forwarding agent, who in turn will pay the transporters. The forwarding agent will take care of selecting a sea carrier, taking into account the project constraints. In some case, the agent will be part of a group that includes a sea carrier, but the two entities shall anyway be separate.

CUSTOM CLEARANCE

The other area where the use of an agent is well justified is the customs and a custom clearance agent can be subcontracted. This is usually a local organization, with a good knowledge of the custom authorities, the procedures, and the "do"s and "don't"s of the custom operations.

A particular aspect of a subcontract with a custom clearance agent is related to the payments. While transportation companies can accept payment terms to be within some days or weeks, or even months, custom authorities need to "see" the money paid first, in order to release the materials. On the other hand, the agent is usually not in a position to advance the money and bear the burden of any negative cash flow. The solution is usually through a zero-balance account, i.e., a bank account opened in the name of the custom clearance agent and money is only deposited by contractor in an amount equal to the duties to be paid when a cargo is expected to arrive. The agent will calculate the amount of the custom duties due and will request contractor to deposit that amount in the account. The money will be solely used for the purpose of paying the corresponding custom duties. In this way, nobody suffers from negative cash flow and above all, there remains no reason for delay in the clearance of the materials solely due to missing payments.

The amount of custom duties must be estimated by contractor at the time of preparing the project budget and submitting a quotation to the owner. Information is available from various sources, but it is always recommended to contact a custom clearance agent during the proposal, to make sure that nothing is overlooked.

A risky area for contractor is the preparation of the paperwork for custom clearance. The risk of delay due to late clearance is already discussed, but there is another risk as well. Some projects are custom exempt. This happens if the local government deems the project has national importance and grants custom exemption benefit to the project owner. In this case, there will be no duties to be paid, but the paperwork must be prepared in a way to satisfy local rules and procedures in conjunction with the exemption. If, due to incorrectness of the paperwork, the owner incurs the payment of duties and/or fines, then this burden must usually be borne by the contractor.

This obligation must of course be stated in the contract and the provision must be carefully negotiated at the time of contract signing.

INCOTERMS

INCOTERMS (INternational COmmercial TERMS) are a series of predefined delivery rules and schemes, conceived to designate common language to all operators. The rules were first published in 1936 and eight versions have been issued so far. Last one is INCOTERMS 2010, published in January 2011.

INCOTERMS 2010 considers 11 schemes, seven applicable to general transportation (i.e., by any means of transportation) and four only relevant to sea and inland water transportation.

The following description applies to each identified rule.

GENERAL TRANSPORTATION

EXW (Ex Works named place of delivery). The seller makes the goods available to be collected at the manufacturer's premises and the purchaser is responsible for all other risks, transportation costs, taxes and duties from that point onwards. This term is commonly used when quoting a price.

FCA (Free Carrier, named place of delivery). The seller gives the goods, cleared for export, to the purchaser's carrier at a specified place. The seller is then responsible for transportation to the specified place of final delivery. This term is commonly used for containers travelling for more than one mode of transport.

CPT (Carriage Paid To, named place of destination). The seller pays to transport the goods to the specified destination. Responsibility for the goods transfers to the purchaser when the seller passes the cargo to the first carrier.

CIP (Carriage and Insurance Paid To, named place of destination). The seller pays for insurance as well as transport to the specified destination. Responsibility for the goods transfers to the purchaser when the seller passes the cargo to the first carrier. CIP is commonly used for goods transported by container for more than one mode of transport. If transportation is only by sea, CIF is often used (see below).

DAT (Delivered at Terminal, named terminal at port or place of destination). The seller pays for transport to a specified terminal at the agreed destination. The purchaser is responsible for the cost of importing the goods. The purchaser takes responsibility once the goods are unloaded at the terminal.

DAP (Delivered at Place, named place of destination). The seller pays for transport to the specified destination, but the purchaser pays the cost of importing the goods. The seller takes responsibility for the goods until they are ready to be unloaded by the purchaser.

DDP (Delivered Duty Paid, named place of destination). The seller is responsible for delivering the goods to the named destination in the purchaser's country, including all costs involved.

SEA AND INLAND WATERWAY TRANSPORT

FAS (Free Alongside Ship, named port of shipment). The seller puts the goods alongside the ship at the specified port where the cargo is going to be shipped from. The seller must get the goods ready for export, but the purchaser is responsible for the cost and risk involved in loading the cargo. This term is commonly used for heavy-lift or bulk cargo (e.g., generators, boats), but not for goods transported in containers for more than one mode of transport (FCA is usually used for this).

FOB (Free on Board, named port of shipment). The seller must get the goods ready for export and load them onto the specified ship. The purchaser and seller share the costs and risks when the goods are on board. This term is not used for goods transported in containers for more than one mode of transport (FCA is usually used for that).

CFR (Cost and Freight, named port of destination). The seller must pay the costs of bringing the goods to the specified port. The purchaser is responsible for risks when the goods are loaded onto the ship.

CIF (Cost, Insurance, and Freight, named port of destination). The seller must pay the costs of bringing the goods to the specified port and also pay for insurance. The purchaser is responsible for risks when the goods are loaded onto the ship.

Figure A8.1 illustrates the same rules in a graphical form.

MODE		From workshop to port of export	Unloading of truck at port of export	Loading on vessel at port	From port of export to port of import	Insurance	Unloading at port of import	Loading on truck at port of import	From port of import to destination	Custom clearance	Custom duties
EXW	Ex works	Buyer	Buyer	Buyer	Buyer	Buyer	Buyer	Buyer	Buyer	Buyer	Buyer
FCA	Free carrier	Seller	Buyer	Buyer	Buyer	Buyer	Buyer	Buyer	Buyer	Buyer	Buyer
CPT	Carriage paid to	Seller	Seller	Seller	Seller	Buyer	Buyer	Buyer	Buyer	Buyer	Buyer
CIP	Carriage and insurance paid to	Seller	Seller	Seller	Seller	Seller	Buyer	Buyer	Buyer	Buyer	Buyer
DAT	Delivered at terminal	Seller	Seller	Seller	Seller	Seller	Seller	Buyer	Buyer	Buyer	Buyer
DAP	Delivered at place	Seller	Seller	Seller	Seller	Seller	Seller	Seller	Seller	Buyer	Buyer
DDP	Delivered duty paid	Seller	Seller	Seller	Seller	Seller	Seller	Seller	Seller	Seller	Seller
FAS	Free alongside ship	Seller	Seller	Buyer	Buyer	Buyer	Buyer	Buyer	Buyer	Buyer	Buyer
FOB	Free on board	Seller	Seller	Seller	Buyer	Buyer	Buyer	Buyer	Buyer	Buyer	Buyer
CFR	Cost and freight	Seller	Seller	Seller	Seller	Buyer	Buyer	Buyer	Buyer	Buyer	Buyer
CIF	Cost insurance and freight	Seller	Seller	Seller	Seller	Seller	Buyer	Buyer	Buyer	Buyer	Buyer

FIGURE A8.1 INCOTERMS delivery scheme.

Glossary of Acronyms

The following is a list of the acronyms and abbreviations used in this book, along with their meaning.

AACE : (formerly) Association for the Advancement of Cost Engineering

ALARP : As low as reasonably practicable

ANSI : American National Standards Institute

AOA : Activity on Arrow

AON : Activity on Node

ASME : American Society of Mechanical Engineers

ASTM : American Society for Testing Materials

ATEX : Atmosphères Explosibles

BAS : Building Automation System

BEDD : Basic Engineering and Design Data

BFD : Block Flow Diagrams

BMS : Building Management System

BNB : Bid-No-Bid

BOQ : Bill of Quantities

BOT : Build–Operate–Transfer

BSI : British Standards Institute

CAD : Computer-Aided Design

CAPEX : Capital Expenditure

CAR : Construction All Risks

CCTV : Closed Circuit Television

CEO : Chief Executive Officer

CEPCI : Chemical Engineering Plant Cost Index

CESCE : Compañia Española de Seguros de Crédito a la Exportación

CFD : Cash Flow Diagram

CFR : Cost and Freight

CIF : Cost, Insurance, and Freight

CIP : Carriage and Insurance Paid

C-LSTK : Converted Lump-Sum Turn-key

COA : Code of Accounts

COFACE : Compagnie Française d' Assurance pour le Commerce Extérieur

CPM : Critical Path Method

CPT : Carriage Paid

DAP : Delivered at Place

DAT : Delivered at Terminal

DCS : Distributed Control System

DDP : Delivered Duty Paid

DG :	Diesel Generator
DIN :	Deutsches Institut für Normung
DSCF :	Debt Service Coverage Factor
E&C :	Engineering and Construction
E&I :	Electrical and Instrumentation
EAR :	Erection All Risks
ECA :	Export Credit Agency
ECGD :	Export Credits Guarantee Department
EFD :	Engineering Flow Diagram
EHS :	Environment, Health, and Safety
EN :	European Standards
ENVID :	Environmental Impact Identification
EP :	Engineering and Procurement
EPC :	Engineering, Procurement, Construction
EPCm :	Engineering, Procurement, Construction management
ES :	Engineering Services
ES :	End-to-Start
ESD :	Emergency Shut Down
ETA :	Event-tree Analysis
EXW :	Ex Works
FAS :	Free Alongside Ship
FCA :	Free Carrier
FCI :	Fixed Capital Investment
FEED :	Front End Engineering & Design
FF :	Flat Face
FF :	Finish-to-Finish
FMEA :	Failure Modes and Effects Analysis
FOB :	Free on Board
FTA :	Fault-tree Analysis
GICS :	Global Industry Classification Standard
HAZID :	Hazard Identification
HAZOP :	Hazard and Operability
HO :	Home Office
HP :	High Pressure
HR :	Human Resources
HSE :	Health, Safety, and Environment
HSSE :	Health, Safety Security, and Environment
HV :	High Voltage
HVAC :	Heating, Ventilation, and Air Conditioning
I/O :	Input/Output
ICB :	Industry Classification Benchmark
ICC :	International Chamber of Commerce
IFC :	Issue for Construction
INCOTERMS :	International Commercial TERMS
IRR :	Internal Rate of Return
ISIC :	International Standard Industrial Classification

ISO :	International Standard Organization
IT :	Information Technology
ITB :	Invitation to Bid
ITT :	Invitation to Tender
KOM :	Kickoff Meeting
KPI :	Key Performance Indicator
L/C :	Letter of Credit
LD :	Liquidated Damages
LLI :	Long Lead Items
LNG :	Liquefied Natural Gas
LOI :	Letter of Intent
LOPA :	Layer of Protection
LP :	Low Pressure
LS :	Lump Sum
LS :	Late Start
LSTK :	Lump-Sum Turn-Key
LV :	Low Voltage
MCC :	Motor Control Center
MD :	Machinery Directive
MFD :	Mechanical Flow Diagram
MOC :	Material of Construction
MOM :	Minutes of Meeting
MP :	Medium Pressure
MR :	Material Requisition
MTO :	Material Takeoff
MV :	Medium Voltage
NAICS :	North American Industry Classification System
NCR :	Non-Conformity Report
NDT :	Non-Destructive Test
NEC :	National Electric Code
NEXI :	Nippon Export and investment Insurance
NFPA :	National Fire Protection Association
NPSH :	Net Pressure on Suction Head
NPV :	Net Present Value
O&M :	Operation and Maintenance
OBE :	Open Book Estimate
OHSAS :	Occupational Health and Safety Series
OPEX :	Operating Costs
OSBL :	Outside Battery Limits
P&ID :	Piping and Instrument Diagram
PBP :	Payback Period
PCG :	Parent Company Guarantee
PDP :	Process Design Package
PDR :	Process Design Revision
PE :	Project Engineer
PEC :	Project Engineering Coordinator

PED :	Pressure Equipment Directive
PEP :	Project Execution Plan
PERT :	Program Evaluation and Review Technique
PFD :	Process Flow Diagrams
PLC :	Programmable Logic Controller
PM :	Project Manager
PMC :	Project Management Contractor
PO :	Purchase Order
PQR :	Procedure Qualification Record
PTW :	Permit to Work
QA :	Quality Assurance
QC :	Quality Control
QM :	Quality Management
QMS :	Quality Management System
QRA :	Quantitative Risk Assessment
R&D :	Research and Development
RF :	Raised Face
RFP :	Requests for Proposal
RFQ :	Request for Quotations
RJ :	Ring Joint
Ro-Ro :	Roll on Roll off
SACE :	Sezione Assicurazione Commercio Estero
SHE :	Safety, Health, and Environment
SIL :	Safety Integrity Level
SIS :	Safety Instrumentation System
SWOT :	Strengths, Weaknesses, Opportunities, and Threats
TCF :	Temporary Construction Facilities
TDA :	Technical Development Allowance
TQM :	Total Quality Management
U/G :	Underground
UID :	Utilities Piping and Instrument Diagrams
UP :	Unit Price
UPS :	Uninterrupted Power Supply
VAT :	Value-Added Tax
WBS :	Work Breakdown Structure
WC :	Working Capital
WPS :	Welding Procedure Specifications

Index

Printed and bound by CPI Group (UK) Ltd, Croydon, CR0 4YY

24/10/2024

01778281-0011